# Why Big Fierce Animals Are Rare

## AN ECOLOGIST'S PERSPECTIVE

# Why Big Fierce Animals Are Rare

## AN ECOLOGIST'S PERSPECTIVE

*Paul Colinvaux*

PRINCETON UNIVERSITY PRESS

PRINCETON, NEW JERSEY

Library of Congress Cataloging in Publication Data will be
found on the last printed page of this book

This book has been composed in V.I.P. Caledonia

Printed in the United States of America


First Princeton Paperback printing, 1979

9  8  7  6

# Contents

# Preface

ECOLOGY is not the science of pollution, nor is it environmental science. Still less is it a science of doom. There is, however, an overwhelming mass of writings claiming that ecology is all of these things. I wrote this book in some anger to retort to this literature with an account of what one practicing ecologist thinks his subject is really about. I feel fervor for the elegance of the Darwinian explanations we find for natural phenomena. I take the opportunity to brand as nonsense tales of destroying the atmosphere, killing lakes and hazarding the world by making it simple.

I wrote a textbook before I wrote this. With the textbook in press, I spent a quiet year as a Guggenheim Fellow thinking out the social implications of ecological knowledge. Then I wrote a series of articles for the *Yale Review* (© Yale University), culminating in a paper describing an ecological model of human history. The pedigree of this book, therefore, is: out of text; by Guggenheim and the *Yale Review*. To the Guggenheim Foundation and the editor of the *Yale Review*, I give thanks.

My own research has been on communities of the past and the history of climatic change as it can be reconstructed from fossils in the mud of ancient lakes. I have inquired into the way of life of the First Americans on the ancient plainsland now drowned under the Bering

Sea and into the environmental history of the Galapagos Archipelago. Most other subjects I report at second hand, though by now many of my summaries have been through the reviewing process of a text in use for several years. This manuscript was reviewed by Dr. R. H. Whittaker and Dr. H. Horn. It would be hard to find two other reviewers who could point out one's errors so nicely without wounding one's feelings. My gratitude to them both. Where I still err it is probably because I was willful in the face of their advice.

To keep my prose as uncluttered as possible I have refrained from references and footnotes. All the main studies and arguments I describe, however, have their sources given in the section at the end called "ecological reading." Most can also be found written up in the half-dozen texts I cite in that section.

I recently had the pleasant task of reading into a tape-recorder from Darwin's autobiography, acting the voice of Darwin for a colleague's production. As I was captured by the part, I could feel the thoughts of that greatest of all ecologists in the room. It is in Darwin's writings that one finds the true roots of ecology. Darwin did not write of pollution and crisis but of how the world worked; of coral reefs and species; of expressing emotions; of fertilizing orchids and natural selection. Ecologists still ponder these things, and in Darwin's way.

**Paul Colinvaux**
*Columbus, Ohio*
*14 February, 1977*

# Why Big Fierce Animals Are Rare

## AN ECOLOGIST'S PERSPECTIVE

# Prelude

THE earth is an object in space, a ball of rock whose hard crust floats on a molten core. The crust writhes and moves with the slow tempo of geologic time, giving rise to that strange pattern of interlocking shapes that are the continents and the great ocean basins. The crust is surrounded by a very thin atmosphere, a curious mixture of the gases oxygen and nitrogen found nowhere else in the solar system, with small vital traces of carbon dioxide and water vapor among the oxygen and nitrogen. This rocky object is flooded with light and heat from our sun, a flow of energy of unremitting fierceness, cascading down upon it.

If you were to look down on the earth from space, all would appear silent and still. The writhings of the earth's crust are too slow to be noticed against the short spans of our lives. Even the rushing motions of the air would be hardly noticeable, partly because of the time scale and the distance involved, and partly because of the transparency of the gases. The only signs of motion are the slowly shifting clouds of water vapor and the change from green to brown to white at high latitudes as summers give way to autumns and winters.

But if you plunge down, close to that rocky crust, into that skin of atmosphere, all is noise and excitement after the everlasting stillness and quietness of space. Not only is there rushing air and pouring water but an extraordinary array of living things murmurs and moves on the face of the earth. These living things are thinly spread where the atmospheric skin clings to the crust of rocks. They share the energy streaming down on the earth, and they share the space of the earth's surface and the vast three-dimensional spaces of the oceans. They exist together in some form of accommodation, living and letting live, always suited to the ways of life they must follow, often present in teemingly diverse array.

The people who study the workings of this array are called "ecologists."

# Chapter One. The Science That Reasons Why

LIFE works from the sun. Solar energy is caught by the plants and used to build forests and prairies and to fuel the life of the sea. But, as fast as it is made, vegetation is always being cut back by the animals that feed upon it, by physical accident, in quarrels among plants, by disease and old age. So the vegetation works to maintain itself, mining nutrients from the soil, always replacing what is lost. We come to look at the working patches of life on earth as series of machines that crank on with self-perpetuating precision, as systems that cycle the raw materials of life, using the fuel the sun provides. Ecologists talk of "ecosystems," and the word is so expressive that its use has spread, even with much of its real meaning, into the common language.

The naturalist's eye discerns the mighty workings of a living system in all places. The ordered array of trees in a temperate forest, the understory bushes in their proper places, the carpet of flowers that bloom in the spring before the trees have spread their leaves, the saplings awaiting their turn in the canopy, the animals eating leaves, spreading pollen, and hoarding nuts, the fiercer animals hunting the first lot down and checking their depredations, the soil underfoot where decomposing remnants of life are stirred by other animals and made to give back their raw materials to be used again by the forest—how well this ecosystem works! And the prairies, marshes, lakes, pine woods, coral reefs, and the Antarctic whaling grounds work just as well. Even meadows and orchards created by people work, and on the same system. We want to know how they work and why they keep on working.

Once we start wondering about these things we discover some remarkable oddities about natural systems that must be explained if we are ever really to understand them. For one thing, the ecosystems have bewilderingly large numbers of moving parts.

When engineers plan systems, they like to keep the number of working parts to an efficient minimum; so an engineered-system looks quite different from an ecosystem. This is shown clearly in the design of spaceships or satellites, which are made to obtain their energy from the sun in the way a forest does. We make gold-plated panels to collect the sun, and we affix them to stalks like golden trees. But because we make only one kind of gold-plated panel, the golden trees on our satellites all look the same. When we learn to make better solar panels, we will scrap the old kind and use the new. But whatever designed the wild forest of green trees found it necessary to make many different kinds, and we shall not understand how an ecosystem of real trees works until we know why all these kinds were necessary.

There is really nothing else so odd about life as its variety. Consider the grasses in a pasture. A dozen kinds of grass and other plants will live all mixed up together in an old pasture. I once found five or six kinds of grass in a square yard or so of one of the manicured lawns of Jesus College, Cambridge (twenty years ago, before the advent of some of the nastier chemicals; but perhaps Jesus College gardeners are still humanist enough to eschew them). But why should there be so many kinds of grass in a pasture or lawn? Why is there not one perfect kind of pasture plant, ideally suited to the circumstances of pasture life, perfectly efficient at getting a living in that grazed or mowed-down space?

Animals live in a pasture too, and in even greater array than the plants, particularly the insects, which are present in dozens of oddly different shapes. Why so many? Why not one perfect grass-processing, pasture insect, together perhaps with its perfectly adapted predator? To understand how the pasture ecosystem goes on working year after year, we certainly need to know why it is important to have so many different kinds of insect.

The same lavish richness is apparent wherever life is found. A hundred thousand kinds of plants are known to science, and estimates of the insect host run upward from a million species. There are eight thousand species of living birds, and other animals in proportion. Why are there so many different kinds of plants and animals? Why, for that matter, are there not more?

Some of the plants and animals are always there, familiar everyday companions, present year after year. But other kinds are rare, or they come upon us only now and then as sudden novelties or plagues. Does the working of nature require the presence of those rare animals and plants, or can we throw them out with impunity? To answer this we had best start by finding out what makes some kinds common and others rare in the first place.

Then there is the problem of constancy of number. In every part of the natural wild plants and animals are breeding as fast as they can, and yet their number seems the same year after year. Our forefathers have been able to tell their children about robins, believing that what they say would always be roughly true, that the robins would be there in after years, friendly birds, common enough but not a plague either. And yet robins, like everything else that lives, are breeding as fast as they can. A robin's nest holds several chicks, and ambitious robins may nest more than once a year, but this does not affect the number of robins. Why do the numbers of every kind of living thing stay so roughly constant? Why do the common stay common and the rare stay rare?

If we do not believe in magic or special creation, then the answers to these questions must be found in the ways the animals and plants get their living from their environment. We know that animals and plants have slowly changed over the years so that the ones we have are beautifully fitted to the lives they must lead. They have evolved to find food, to survive hazard, and to make babies under the local circumstance or "environment." If we are to understand the lives they lead, why there are so many of them, and why there are not more, we must study the environment itself, the resources the animals and plants need, and what they must do to get these resources.

And so an ecologist who begins by asking himself how life works in those splendid perpetual systems very quickly finds himself asking instead why the natural ecosystems are made up of so many parts and why there are so many of each kind of part. Before he can answer an engineer's question "how does this work," he is faced with even more fundamental questions beginning with "why": Why are there so many different kinds of plants and animals? Why are some common but others rare? Why are some large but others small? Why do they sometimes do such peculiar things?

We approach our dual set of questions with the knowledge that all the machinations of life in ecosystems must be products of the process of natural selection. Species change and have been changing for more than a thousand million years. Of this we are as certain as of anything in science. The different kinds are constantly fashioned by a mindless selective force acting always to destroy what is least suitable, letting live what is most suited or "fit."

Natural selection designs species. It never invents a design; it merely chooses from the range of varieties that happens to be at hand. By this act of choice, however, it sets the existing designs of species all the same. Since the workings of ecosystems must result from the doings of their individual parts, it follows that if we would understand the engineer's questions of how life works we must at the same time understand why the outcome of natural selection is the actual array of species we find in any particular place.

Ecologists are growing more confident that they can answer many of these vital questions. They think they know why some animals are common and others rare, why some are bigger than others, why their numbers are the same year after year, why their behavior may be curious, and how they share the life-giving energy that comes from the sun. In this book I try to trace the status of the ecologist's quest, concentrating on some of the more exciting and hard-fought intellectual struggles.

## Chapter Two.  Every Species Has Its Niche

EVERY species has its niche, its place in the grand scheme of things.

Consider a wolf-spider as it hunts through the litter of leaves on the woodland floor. It must be a splendid hunter; that goes without saying for otherwise its line would long since have died out. But it must be proficient at other things too. Even as it hunts, it must keep some of its eight eyes on the look-out for the things that hunt it; and when it sees an enemy it must do the right thing to save itself. It must know what to do when it rains. It must have a life style that enables it to survive the winter. It must rest safely when the time is not apt for hunting. And there comes a season of the year when the spiders, as it were, feel the sap rising in their eight legs. The male must respond by going to look for a female spider, and when he finds her, he must convince her that he is not merely something to eat—yet. And she, in the fullness of time, must carry an egg-sack as she goes about her hunting, and later must let the babies ride on her back. They, in turn, must learn the various forms of fending for themselves as they go through the different moults of the spider's life until they, too, are swift-running, pouncing hunters of the woodland floor.

Wolf-spidering is a complex job, not something to be undertaken by an amateur. We might say that there is a profession of wolf-spidering. It is necessary to be good at all its manifold tasks to survive at it. What is more, the profession is possible only in very restricted circumstances. A woodland floor is necessary, for instance, and the right climate with a winter roughly like that your ancestors were used to; and enough of the right sorts of things to hunt; and the right shelter when you need it; and the numbers of natural enemies must be kept within reasonable bounds. For success, individual spiders must be superlatively good at their jobs and the right circumstances must prevail. Unless both the skills of spidering and the opportunity are present, there will not be any wolf-spiders. The "niche" of wolf-spidering will not be filled.

"Niche" is a word ecologists have borrowed from church architecture. In a church, of course, "niche" means a recess in the wall in which a figurine may be placed; it is an address, a location, a physical place. But the ecologist's "niche" is more than just a physical place: it is a place in the grand scheme of things. The niche is an animal's (or a plant's) profession. The niche of the wolf-spider is everything it does to get its food and raise its babies. To be able to do these things it must relate properly to the place where it lives and to the other inhabitants of that place. Everything the species does to survive and stay "fit" in the Darwinian sense is its niche.

The physical living place in an ecologist's jargon is called the *habitat*. The habitat is the "address" or location" in which individuals of a species live. The woodland floor hunted by the wolf-spiders is the habitat, but wolf-spidering is their niche. It is the niche of wolf-spidering that has been fashioned by natural selection.

The idea of "niche" at once gives us a handle to one of those general questions that ecologists want to answer —the question of the constancy of numbers. The common stay common, and the rare stay rare, because the opportunities for each niche, or profession, are set by circumstance. Wolf-spidering needs the right sort of neighbors living in the right sort of wood, and the number of times that this combination comes up in any country is limited. So the number of wolf-spiders is limited also; the number was fixed when the niche was adopted. This number is likely to stay constant until something drastic happens to change the face of the country.

Likening an animal's niche to a human profession makes this idea of limits to number very clear. Let us take the profession of professing. There can only be as many professors in any city as there are teaching and scholarship jobs for professors to do. If the local university turns out more research scholars than there are professing jobs, then some of these hopeful young people will not be able to accept the scholar's tenure, however *cum laude* their degrees. They will have to emigrate, or take to honest work to make a living.

Likewise there cannot be more wolf-spiders than there are wolf-spider jobs, antelopes than there are antelope jobs, crab grass than there are crab grass jobs. Every species has its niche. And once its niche is fixed by natural selection, so also are its numbers fixed.

This idea of niche gets at the numbers problem without any discussion of breeding effort. Indeed, it shows that the way an animal breeds has very little to do with how many of it there are. This is a very strange idea to someone new to it, and it needs to be thought about carefully. *The reproductive effort makes no difference to the eventual size of the population.* Numerous eggs may increase numbers in the short term, following some disaster, but only for a while. The numbers that may live are set by the number of niche-spaces (jobs) in the environment, and these are quite independent of how fast a species makes babies.

But all the same each individual must try to breed as fast as it can. It is in a race with its neighbors of the same kind, a race that will decide whose babies will fill the niche-space jobs of the next generation. The actual number of those who will be able to live in that next generation has been fixed by the environment; we may say that the population will be a function of the *carrying capacity* of the land for animals of this kind in that time and place. But the issue of whose babies will take up those limited places is absolutely open. It is here that natural selection operates. A "fit" individual is, by definition, one that successfully takes up one of the niche-spaces from the limited pool, and the fitness of a parent is measured by how many future niche-spaces her or his offspring take up. "Survival of the fittest" means survival of those who leave the most living descendants. A massive breeding effort makes no difference to the future population, but it is vital for the hereditary future of one's own line. This is why everything that lives has the capacity for large families.

Yet there are degrees of largeness in wild families, and these degrees of largeness make sense when looked at with an ecologist's eye. The intuitively obvious consequence of a law that says "Have the largest possible family or face hereditary oblivion," is the family based on thousands of tiny eggs or seeds. This seems to be the commonest breeding strategy. Houseflies, mosquitoes, salmon, and dandelions all do it. I call it "the small-egg gambit." It has very obvious advantages, but there are also costs, which the clever ones with big babies avoid.

For users of the small-egg gambit, natural selection starts doing the obvious sums. If an egg is made just a little bit smaller, the parent might be able to make an extra egg for the same amount of food eaten, and this will give it a slight edge in the evolutionary race. It is enough. Natural selection will therefore choose families that make more and more of smaller and smaller eggs until a point of optimum smallness is reached. If the eggs are any smaller than this, the young may all die; if they are any larger, one's neighbor will swamp one's posterity with her mass-production. The largest number of the smallest possible eggs makes simple Darwinian sense.

But the costs of the small-egg gambit are grim. An inevitable consequence is that babies are thrown out into the world naked and tiny. Most of them as a result die, and early death is the common lot of baby salmon, dandelions, and the rest. In the days before Darwin, people used to say that the vast families of salmon, dandelions, and insects were compensations for the slaughter of the young. So terrible was the life of a baby fish that Providence provided a salmon with thousands of eggs to give it a chance that one or two might get through. It seems a natural assumption, and one that still confuses even some biologists. But the argument is the wrong way round. A high death rate for the tiny, helpless young is a consequence of the thousands of tiny eggs, not a cause. A selfish race of neighbor against neighbor leads to those thousands of tiny eggs, and the early deaths of the babies are the cost of this selfishness.

There is this to be said for the small-egg gambit, though; once you have been forced into it, there are the gambler's compensations. Many young scattered far and wide mean an intensive search for opportunity, and this may pay off when opportunity is thinly scattered in space. Weed and plague species win this advantage, as when the parachute seed of a dandelion is wafted between the trunks of the trees of a forest to alight on the fresh-turned earth of a rabbit burrow. The small-egg gambits of weeds may be likened to the tactics of a gambler at a casino who covers every number with a low-value chip. If he has enough chips, he is bound to win, particularly if big payoffs are possible. He does have to have very many chips to waste, though. This is why economists do not approve of gamblers.

To the person with an economic turn of mind, the small-egg gambit, for all its crazy logic, does not seem a proper way to manage affairs. The adherents of this gambit spend all their lives at their professions, winning as many resources as possible from their living places, and then they invest these resources in tiny babies, most of whom are going to die. What a ridiculously low return on capital. What economic folly. Any economist could tell these animals and plants that the real way to win in the hereditary stakes is to put all your capital into a lesser number of big strong babies, all of which are going to survive. A number of animals in fact do this. I call it "the large-young gambit."

In the large-young gambit one either makes a few huge eggs out of the food available, or the babies actually grow inside their mother, where they are safe. Either way, each baby has a very good chance of living to grow up. It is big to start with and it is fed or defended by parents until it can look after itself. Most of the food the parents collect goes into babies who live. There is little waste. Natural selection approves of this as much as do economists. Big babies who have a very good chance of long life mean more surviving offspring for the amount of food-investment in the end. This prudent outlay of resources is arranged by birds, viviparous snakes, great white sharks, goats, tigers, and people.

Having a few, large young, and then nursing them until they are big and strong, is the surest existing method of populating the future. Yet the success of this gambit assumes one essential condition. You must start with just the right number of young. If you lay too many monster hen's eggs or drop too many bawling brats, you may not be able to supply them with enough food, and some or all will die. You have then committed the economic wastefulness of those of the tiny eggs. So you must not be too ambitious in your breeding. But the abstemious will also lose out, because its neighbor may raise one more baby, may populate the future just that little bit better, and start your line on a one-way ride to hereditary oblivion. You must get it just right; not too many young, and not too few. Natural selection will preserve those family strains which are programmed to "choose" the best or optimum size of family.

Many ecologists have studied birds with these ideas in mind, and they have found that there is often a very good correlation between the number of eggs in a clutch and the food supply. In a year when food is plentiful a bird may lay, on the average, one or two eggs more than in a lean season. The trend may be slight but sometimes is quite obvious. Snowy owls, which are big white birds of the arctic tundra, build vast nests on the ground. They feed their chicks on lemmings, the small brown arctic mice. When lemmings are scarce, there may be only one or two eggs in each owl's nest, but when the tundra is crawling with lemmings, the nests may well have ten eggs each. The owls are evidently clever at assessing how many chicks they can afford each year.

But people are cleverer than snowy owls and have brought the large-young gambit to its perfection. They can read the environment, guess the future, and plan their families according to what their intelligence tells them they can afford. Even the infanticide practiced by various peoples at various times serves the cause of Darwinian fitness, rather than acting as a curb on population. There is no point in keeping alive babies who could not be supported for long. Killing babies who could not be safely reared gives a better chance of survival to those who are left, and infanticide in hard times can mean that more children grow up in the end.

Thus, every species has its niche, its place in the grand scheme of things; and every species has a breeding strategy refined by natural selection to leave the largest possible number of surviving offspring. The requirement for a definite niche implies a limit to the size of the population because the numbers of the animal or plant are set by the opportunities for carrying on life in that niche. The kind of breeding strategy, on the other hand, has no effect on the size of the usual population, and the drive to breed is a struggle to decide which family strains have the privilege of taking up the limited numbers of opportunities for life. Every family tries to outbreed every other, though the total numbers of their kind remain the same. These are the principles on which an ecologist can base his efforts to answer the major questions of his discipline.

# Chapter Three. Why Big Fierce Animals Are Rare

ANIMALS come in different sizes, and the little ones are much more common than the big.

A typical small patch of woodland in any of the temperate lands of the North will contain hosts of insects and then nothing larger running about until we get to the size of small birds, which are much less numerous. Another size jump brings us to foxes, hawks, and owls, of which there may be only one or two. A fox is ten times the size of a song bird, which is ten times the size of an insect. If the insect is one of the predacious ground beetles of the forest floor, which hunt among the leaves like the wolf-spiders, then it, in turn, is ten times bigger than the mites and other tiny things that they both hunt.

The animals in this system of living do indeed come in very distinct sizes. There are, of course, some in-between ones, but not many. Squirrels in the upper size range seem obvious, but I am hard put to find something between an insect and a small bird unless it is a newt or lizard, neither of them very prominent denizens of a temperate woodland. Slugs and snails are toward the size of caterpillars. Shrews and toads are near the size of song birds. Even a snake can be thought of as an odd-shaped hawk.

In the wood as elsewhere there are distinctly different sizes, and the little ones are the most common. The same sort of thing exists in the sea in even odder form, for in the open sea the really tiny things are plants; the microscopic diatoms and other algae. Ten times bigger than these (give or take a few times) are the animals of the plankton, the copepods and the like. Bigger still are the shrimps and fish that hunt those copepods. Then another jump brings us to herrings, then to sharks, or killer whales. In any one place in the sea, this clumping of life into different sizes is the normal thing.

In the sea the rarity of the large is also most clearly shown. Great white sharks are extremely rare, and the other kinds of shark are scattered pretty thinly over the seas too. Fish of the herring size are vastly more common than sharks, but, even so, the number that are seen in a casual dive in the sea is seldom immense. If you drift, and focus your eyes just outside the facemask, however, myriad darting specks of the smaller animals may become visible. If you later take some of that same water and spin it in a centrifuge, there is likely to be a thin green scum in the bottom made up of an almost uncountable multitude of independent, tiny plants.

The tiny things of woodland and sea are immensely common; bigger things are a whole jump bigger and a whole jump less common; and so on until we reach the largest and rarest animals of all. A like pattern can be found in tropical forests, Irish bogs, or just about anywhere else. It is an extraordinary thing but true that life comes in size-fractions which, for all the blending and exceptions that can be found by careful scrutiny, are remarkably distinct. Animals in the larger sizes are comparatively rare.

Charles Elton of Oxford pointed out this strange reality half a century ago. Elton went adventuring on Spitzbergen, an arctic island covered with treeless tundra, where the animals move about in the open and where particularly he could follow an arctic fox as it went about its daily affairs. Arctic foxes can be delightfully tame. On St. George Island in the Bering Sea one tried to take sandwiches from my pocket as I sat upon a rock. Elton followed his foxes and pondered their activities through a summer that was to be one of the most important an ecologist ever spent.

The foxes caught the summer birds of the tundra—the ptarmigan, sandpipers, and buntings; and these birds were at once a size-jump smaller than the foxes and much more numerous. The ptarmigan ate the fruit and leaves of tundra plants, but the sandpipers and buntings ate insects and worms, which were again a size-jump smaller as well as being more numerous. The foxes also ate seagulls and eider ducks, smaller and more numerous than the foxes, and these birds ate the tiny abundant life of the sea. Elton not only saw all this but, as Sherlock Holmes often lectured Watson, he *observed* it also. That small things were common and large things rare has been known by everybody since the dawn of thought, but Elton pondered it as Newton once pondered a falling apple, and knew he was watching something odd. Why should large animals be so remarkably rare? And why should life come in discrete sizes?

Elton's summer on Spitzbergen gave him the answer to the second of these questions even as he posed it. The discrete sizes came about from the mechanics of eating and being eaten. He had seen a fox eat a sandpiper and a sandpiper eat a worm. These animals of different sizes were linked together by invisible chains of eating and being eaten. Foxes had to be big enough, and active enough, to catch and eat the birds on which they preyed; and the birds likewise must overpower, and engulf at a single swallow, the animals on which they fed. The normal lot of an animal was to be big enough to vanquish its living food with ease, and usually to be able to stuff it down its throat whole or nearly so. As one moves from link to link of a food chain the animals got roughly ten times bigger. Life comes in discrete sizes because each kind must evolve to be much bigger than the thing it eats.

Elton's conclusions were obviously true in a very general way. The communities of woodland and ocean on which his thinking was based seemed to conform very nicely. Life in those communities did come in different sizes, and it seemed that the sizes had grown discrete because each kind had evolved to be much bigger than the thing it ate. But many exceptions to the general principle of food size come to mind: wolves, lions, internal parasites, elephants, and baleen whales. There are many animals that are either smaller than their food, such as wolves or parasites, or else absurdly bigger like whales. But a closer look at any of these animals shows them to be instructive exceptions, if true exceptions to the rule all the same.

Land herbivores do not fit the Eltonian model, at least not completely, because land plants provide different-sized mouthfuls for different sizes of animals. You do not have to kill an entire land plant in order to eat it, you just tear off a suitable piece, a shoot, some grass blades, a berry, a bite out of a leaf. Food chains based on vegetation could start with many different sizes of plant-eating animal because squirrels, caterpillars, and elephants share the same food. Even so there does not seem to be a complete continuum of sizes amongst vegetarians, at least in any one place. Both big and little plant-eaters exist in a forest or a prairie and there is not much difficulty in sorting them into sizes. This is because the predators of plant-eaters do have to be size-conscious when they look for food. A selection pressure acts downward along the food-chains, as herbivores evolve sizes that let them escape even as carnivores evolve sizes that enable them to catch skillfully. It is as important to be of a size that does not fit in someone else's mouth as it is to have a mouth suited to the size of one's own prey. So natural selection tends to preserve size classes even when food chains start with a pablum of meadow-forage or forest.

Wolves sometimes obey Eltonian principles, as when they hunt singly for rodents and small game, but they have evolved the trick of packing-up to haul down bigger prey, in winter, when they are freed from family cares and can go out in gangs. Other pack-hunting animals work variants on this method. And all large carnivores have had their sizes adjusted to the needs of killing, rather than of engulfing, so that a lion needs to be big enough to pull down an ailing zebra, but no bigger.

Parasites are smaller than their food, for obvious reasons, but their activities still tend to separate the animals on parasite food chains into different sizes with every link as was described before the coming of ecology in a jingle by Jonathan Swift:

> Big fleas have little fleas
>   Upon their backs to bite 'em
> And little fleas have lesser fleas
>   And so *ad infinitem.*

Very special sea animals such as whales are even more instructive, and we discuss them at the end of this chapter. But otherwise in the sea the pattern of size tends to conform very well to the simplest interpretation of the workings of food size. This is because the sea plants are tiny, individual, and have to be hunted and killed by those who would live off them (seaweeds of the coasts are of trivial importance in the wide oceans). So in the sea a rather complete set of steps runs up the food chains from the smallest plants, through crustaceans and fish, to great white sharks.

Thinking these Eltonian thoughts brings up another of nature's conundrums, "Why are land plants big but sea plants small?" But that must wait for another chapter.

Now there was the matter of rarity. Elton showed that there had to be size-jumps as one went up food chains, and that the animals on the upper end had to be big. But why should the big be so rare? And very rare they are. One has only to compare the number of sharks to the number of herrings, or warblers to caterpillars, to see this. With every jump in size an even mightier loss occurs in numbers. Elton coined a term to describe this fact of life; he called it "The Pyramid of Numbers." He saw in his mind's eye a mighty host of tiny animals supporting on their backs a much smaller army of animals ten times as big. And this array supported, in turn, other animals ten times bigger still, but these were a select few. It was a graph of life he imagined with numbers of individuals along the horizontal axis, and position in the chain of eating, together with size, on the vertical. His vision saw the functioning of animal communities like the profile of Zoser's step pyramid at Saqqara, a triangular edifice built of stacked square-ended layers so that the summit could be reached by four or five giant steps. When ecologists forgather they call this result the "Eltonian Pyramid." Now, why should there be pyramids of numbers in nature wherever we look, from the arctic tundras to tropical forests and the open spaces of the sea? Why should large animals, particularly large hunting animals, always be so amazingly rare?

It is tempting to say that no problem exists, that it stands to reason that there cannot be as many big things as little. But this claim suggests that the Eltonian pyramid reflects no more than the elementary facts of spatial geometry. There is clearly no shortage of actual space to hold more big animals. On Spitzbergen, for instance, each fox had acres and acres to run around in, and the world oceans could hold mind-boggling quantities of the large sharks and killer whales who are the top carnivores of the sea. Large plants are crammed together on the earth in astounding numbers so that we call the result a "forest." Only the large animals are discriminated against.

A second tempting argument is to say that there is a finite amount of flesh (what ecologists call "biomass") to go round and that this chunk of flesh could be used either to make a few big bodies or to make very many little ones. The big are rare because they take large slices from their cake. This assertion is true as far as it goes, but it does not go nearly far enough. If, instead of counting the animals in the different size levels of the pyramid, one weighed them, one finds that there is vastly more flesh in the smaller classes, a greater standing crop of life as well as more numerous individuals. All the insects in a woodlot weigh many times as much as all the birds; and all the songbirds, squirrels, and mice combined weigh vastly more than all the foxes, hawks, and owls combined. The pyramid of numbers is also a pyramid of mass, and the problem remains unsolved. Why is there so little living tissue in the larger animal sizes?

Elton did not have the answer. He thought it might be because little animals reproduced very quickly (true, they do—compare the egg output of butterflies with that of the birds that eat caterpillars) and that rapid reproduction was the key to vast populations. But this is to fall into the age-old error of biologists and theologians alike, the error that says numbers are set by breeding strategy. We have discussed this non-Darwinian idea in the last chapter. Numbers are set by the opportunities for one's way of life, not by the way one breeds. Professorships set the limit to the population of professors, not the productive output of graduate schools. The fact that large animals are rare cannot have anything to do with their reproductive drives. Elton's explanation will not do.

It took nearly twenty years for the corporate body of science to come up with the answer to the question Elton posed in 1927. Raymond Lindeman and Evelyn Hutchinson did so at Yale by thinking of food and bodies as calories rather than as flesh.

A unit of biomass or flesh represents a unit of potential energy that is measured in calories. If we burn a chunk of protein we liberate so many calories of heat, and if we burn a chunk of fat we get more calories still. This is now common knowledge to the affluent peoples of the West who worry about the calories in their food lest they become obese. In the 1930s and 1940s even illiterate Hollywood starlets knew this, but biologists wakened to the idea of the calorie rather more slowly. Yet in the use of food as calories lay the answer to the rarity of the large and fierce.

Measuring an animal's flesh in calories also alerts one's mind to the vital fact that bodies represent fuel as well as vessels for the soul. An animal continually burns up its fuel supply to do the work of living, puffing the exhaust gases out of the smokestacks of its mouth and nostrils and sending the calories off to outer space as radiant heat. The animal uses up its flesh, replacing the lost substance by eating more food, then burning most of this up too. This process of consuming matter by the fires of life goes on in every level of the Eltonian pyramids, and the fires are continually fresh-stoked by the plants on which the animal pyramids rest. At each successive level in the pyramids, the animals have to make do with the fuel (food) that can be extorted from the level below. But they can only extort some fraction of what the level below had not itself used up, and with this tithe the denizens of the upper layers must both make their own bodies and fuel their lives. Which is why their numbers are only a fraction of the numbers below, which is to say why they are rare.

The ultimate furnace of life is the sun, streaming down calories of heat with never-fainting ray. On every usable scrap of the earth's surface a plant is staked out to catch the light, its green array of energy receptors and transducers tuned and directed to the glowing source like the gold-plated cells on the arms of a satellite. In those green transducers we call leaves, the plants synthesize fuel, taking a constant allotment of the streaming energy of the sun. Some of this fuel they use to build their bodies, but some they burn to do the work of living. Animals eat those plants, but they do not get all the plant tissue, as we know because the earth is carpeted brown with rotting debris that has not been part of an animal's dinner. Nor can the animals ever get the fuel the plants have already burned. So there cannot be as much animal flesh on the earth as there is plant flesh. It is possible for large plants to be vastly abundant and ranked side by side, but animals of the same size would have to be thinly spread out because they can only be a tenth as abundant.

This would be true even if all animals were vegetarian. But they are not. For flesh eaters, the largest possible supply of food calories they can obtain is a fraction of the bodies of their plant-eating prey, and they must use this fraction both to make bodies and as a fuel supply. Moreover their bodies must be the big active bodies that let them hunt for a living. If one is higher still on the food chain, an eater of a flesh-eater's flesh, one has yet a smaller fraction to support even bigger and fiercer bodies. Which is why large fierce animals are so astonishingly (or pleasingly) rare.

Thus was the grandest pattern of rarity and abundance in the world explained by two men at Yale in the 1940s. Ways of life were bumping against that most fundamental of physical restraints, the supply of energy.

As the realization of what Lindeman and Hutchinson had done for natural history percolated through the consciousness of biology in the 'fifties and 'sixties a thrill of self-respect began to throb in its younger practitioners. Here the pattern of field experience was linked to the fundamental laws of physics. We were talking of energy degraded step by step as it flowed down food chains, losing its power to do work and pouring steadily away to the sink of heat. The grand pattern of life that Elton had seen on Spitzbergen and that countless naturalists had intuitively known before was clearly and directly a consequence of the second law of thermodynamics.

We can now understand why there are not fiercer dragons on the earth than there are; it is because the energy supply will not stretch to the support of superdragons. Great white sharks or killer whales in the sea, and lions and tigers on the land, are apparently the most formidable animals the contemporary earth can support. Even these are very thinly spread. One may swim many lifetimes in the world oceans without encountering a great white shark; and an ancient Chinese proverb asserts that a hill shelters only one tiger. Evolutionary principle tells us that the existence of these animals creates a theoretical possibility for other animals to evolve to eat them, but the food calories to be won from the careers or niches of hunting great white sharks and tigers are too few to support a minimum population of animals as large and horribly ferocious as these would have to be. Such animals, therefore, have never evolved. Great white sharks and tigers represent the largest predators that the laws of physics allow the contemporary earth to support.

But here we run into what seems to be the first real difficulty of the argument. There are living animals that are much larger than tigers and sharks, and there have been some very big ones in the past. How does their existence square with our interpretation of the second law of thermodynamics?

Elephants and the big, cloven-hoofed animals are larger than tigers. In the past there have been even bigger mammals, such as giant ground-sloths and *Titanotherium*, a beast like an overgrown elephant and the largest land mammal ever. There have also been the largest reptiles of the Mesozoic, the ponderous dinosaurs: *Stegosaurus, Brontosaurus, Iguanadon*. None of these animals poses any difficulty for the model. They have all been plant-eaters. In the strict Eltonian model the plant-eaters are small, and indeed in life most of them actually are. In the open sea this rule that plant-eaters must be small is strictly enforced because the drifting plants are so tiny that only very small animals can make a successful living by eating them. But on land, plants often appear as continuous mats of leaves, which we call vegetation, and it is possible for enormous sluggish animals to slurp them up without much nicety in the hunting. Masses of energy are available in the plant-eating niches at the bottom of the Eltonian pyramids, with the result that viable populations of even enormous animals can be supported. The brontosaur and the elephant alike, therefore, leave both our belief in the energy-flow model and the second law of thermodynamics intact.

That leaves two trickier kinds of animals to explain away: the great baleen whales of the contemporary oceans, which are the largest animals ever to have lived, and the flesh-eating dinosaurs such as *Tyranosaurus rex*. These are both meat-eating animals, and they are impressively bigger than great white sharks or tigers.

The baleen whales have learned to cheat, hunting their food in non-Eltonian ways. Essential to the normal structure of the Eltonian pyramid was that every carnivorous animal should have a direct relationship to the size of its food, being big enough to catch and eat it but not so big that the food item should prove a trivial mouthful not worth the effort of hunting. On this model, the food of a blue or right whale should be several feet long. But it is not. The whales cheat with their sieves of baleen, which let them strain from the surface of the sea the tiny shrimps called krill in huge numbers and with little effort. The whales have cut out the middlemen, avoiding all the energy losses that would have accrued if the krill had been passed to a fish and that fish passed to a bigger fish before the whale had its chance at it like any other Eltonian feeder. So the whales, although not plant eaters, feed very low on food chains where the energy supply is still comparatively large. Floating as they do in the sea, they use little energy in their sluggish hunting, paddling quietly along with their mouths open, straining the meat out of the oceanic soup. So the apparent exception of the whales is no exception at all, and our model may stand.

*Tyranosaurus rex* is more difficult for the argument. Tyranosaurs were huge carnivorous dinosaurs, often pictured as a great green kangaroo-like form with a hideous toad-like head, nightmare teeth, and a pair of useless little flapping arms dangling below the ugly neck. An animal answering to the name of *Tyranosaurus* of this size certainly existed, for we have specimens of all his bones. He was several times larger than lions or tigers, or indeed of any other recorded predator. What enabled it to escape the constraints apparently placed on all its successors by the second law of thermodynamics?

It is useful to note first that the tyranosaur fed at the same level as its modern successors, the big cats, and at the same level as the baleen whales in the sea. It fed on plant-eaters relatively low in the food chain, close to the bottom of the Eltonian pyramid, where there was still much energy to be won. A large body, therefore, does not seem hopelessly out of the question. We know that there were many kinds of very large herbivores about in the tyranosaur's time, animals that, in the absence of pack-hunting predators such as dogs, could be overcome only by very powerful attackers. So we might conclude that the necessity for Mesozoic predators to be large and ferociously active is self-evident. There was nothing else to get at the meat so massively on the hoof, so natural selection provided *Tyranosaurus rex*.

I have always been unhappy about this reasoning. If natural selection could fashion a tyranosaur at that time, why not in all subsequent time? Why in particular was there nothing like a tyranosaur in the great age of mammals, that later part of the Tertiary epoch when all the plainslands of the earth held herds of game that make the herds of modern Africa seem trivial by comparison? I have felt compelled to conclude that the constraints on the size of ferociously active predators that has been applied throughout the age of mammals ought to have applied to the reptiles of the Mesozoic era also. By thinking thus I maneuver myself into the position of saying that, on ecological grounds, the *Tyranosaurus rex* did not exist. And yet there the bones are, indubitably the bones of a large flesh-eating animal of the size claimed. It was with a sense of inward peace that I saw a drawing of a recent attempt to put the bones together differently.

The classic picture of the hopping, predacious tyrant-lizard is derived from nineteenth-century reconstructions of the animal. The new reconstruction, first published in *Nature* in 1968, shows the animal to be a waddling, slow-moving beast, not at all the sort one can imagine dashing after a herd of galloping brontosauri. But it probably got them all the same, picking out the sick and the dying, often getting them only as carrion. The tyranosaur was not a ferociously active predator. It did not stand upright, nor did it hop. It held that massive body horizontally, perhaps able to move swiftly for short periods as it balanced its motion with the long tail. But most of its days were spent lying on its belly, a prostration that conserved energy and from which it periodically roused itself, lifting its great bulk on those two little arms in front until it could balance on the thick walking legs. The tyranosaur did indeed support a large mass by meat-eating, but it escaped the energy-consuming price of being active in order to overcome prime specimens of the giant prey it ate. It managed on land essentially the same stratagem that the baleen whales managed in the sea; it found a non-Eltonian way of getting the meat of plant-eaters without having to hunt them properly. Nothing like it has been seen since because the true active predators of the age of mammals were able to clean up the meat supplies before a sluggish beast such as a tyranosaur could get to them. And active predators might even have eaten the tyranosaur itself.

*Tyranosaurus rex*, as popularly portrayed, is a myth. But it is probably safe to say that it will be as durable as any other myth in our culture. The size and ferocity of real-life predators is restricted to the scale of a tiger, and even these must always be rare. The second law of thermodynamics says so.

# Chapter Four. The Efficiency of Life

WE now have a scientific overview of how an ecosystem works. Green plants share out the space available to the ecosystem among themselves and on professional lines. Each kind of plant has a separate niche, specializing in living on good soil or bad, being early in the season or late, being big or little. And these green plants trap some of the energy of the sun to make fuel. Some of this fuel they use, some is taken by animals, much goes to rot. The fuel taken by the animals at the bottom of the Eltonian pyramid is mostly burned up by the herbivores themselves, but a portion is taken by their predators, and so on for one or two more links up the food chains. At each level in the pyramid there are many species of animals, the numbers of each being set by its chosen profession or niche. All the animals and plants use much of their fuel to make as many babies as possible, and many of these babies are used as fuel by other animals. Every animal and plant in this ecosystem has an appointed place defined both by its level in the pyramid and by its niche. All these living things are tied together in a great web of eating and being eaten, and an ecosystem is a complex community of energy-consumers, all straining to get the most and do their best with it. The result of all these individual efforts is the self-perpetuating mechanism of nature at which we wonder.

But how good is that mechanism really? It certainly

works, and it undoubtedly is long-lasting, but is it efficient? This question has more than academic interests
because the future of our human population depends on
the fuel-gathering efficiencies of ecosystems. So we ask
whether the plants and animals of wild ecosystems are
efficient converters of energy, and whether the agricultural ecosystems on which we depend are better or
worse than the wild ones. Once we know the answers to
these questions, we want to know what sets the limits to
efficiency and whether we can do anything to improve
upon whatever it is. We first look at the plants, because
they perform the most important task of subverting the
sun to make fuel, and ask how efficient they are as factories of fuel.

The plants that now exist must be "fit" plants, they
must be able to leave more offspring than have plants
that might have been, which in turn means that they
must be able to win more food than could the might-
have-beens, which means that they must be more efficient at trapping the sun than were the might-have-
beens. Thus a Darwinian ecologist expects all plants to
be superbly efficient. We see that the green receptors
and transducers of energy that we call "leaves" are indeed stacked up on the face of the earth in formidable
array. So far so good. But we expect the chemistry and
thermodynamics of those green transducers to be as efficient as the leaves are abundant. We hear engineers talk
about the efficiency of automobiles or steam engines, by
which they mean how much of the energy supplied as
fuel is converted to useful work. They often talk of
efficiencies of 20 or 30 percent. With these thoughts, we
turn to practical measurement of what plants and animals can really do.

The efficiency of plants was first determined by a fine
piece of armchair scholarship. It was done by Nelson

Transeau in an office of an old building of The Ohio State University in Columbus when he was seeking material for a presidential address to the local academy of sciences. The plant on which this scholar mused was the humble corn plant, so suitable for armchair scholarship because anything measurable about corn can be found out from the library. No one had thought before how to measure its efficiency, but they had measured everything an ingenious man might need to calculate it.

A crop of corn begins with bare, ploughed ground, a place of zero production, zero efficiency. The corn then grows, zealously defended by the farmer from browsing animals and pests, until maturity. During the intervening weeks the corn plants have been receiving sunlight and converting it first to sugar, then to all the other ingredients of the plant's structure. Every calorie these corn plants trapped had one of two possible fates: either it was burned by the plant itself to do the work of growing and living or it was still there at harvest time, dormant as potential energy in that standing crop. Corn plants have been weighed often enough, and an agricultural handbook readily gives average figures for yield of grain, leaves, stem, roots, everything. Also known is how many calories are in a gram of grain, leaves, roots and the rest; just as the amount of calories in a gram of sugar or ice cream is known. So one can add up the calories in a field of corn. Finding out how many calories the plants have burned during their lives is more tricky, but, as we shall see, this can be discovered too.

Transeau mused about an acre of land in the state of Illinois, a good place to begin because someone had measured how many calories came onto the land of that state from the sun on a typical summer's day. A nice crop of good corn growing on that acre would constitute a population of ten thousand plants. These grew from germination to harvest, as it happened, in exactly one hundred days. Now it was necessary only to go to the handbooks to find out how much poundage was represented by ten thousand well-grown corn plants. Transeau did this, then did a little calculation to convert all the cellulose, protein, and other chemicals they represented back into the sugar from which they had originally been made. In his mind's eye, Transeau saw not a field of ten thousand yellowing, rustling plants but a beautiful pile of glistening white sugar. The sugar weighed 6,678 kilograms.

Now Transeau needed only to know how much sugar these ten thousand plants had burned in their hundred days of life, and his own notebooks gave him this figure. Transeau had pioneered the measurement of breathing in plants, and by the time of that presidential address of 1926 he had all the figures he needed. These had come from corn plants that Transeau had grown in glass chambers to which he could control the air supply. He measured the carbon dioxide going into the chambers and the carbon dioxide coming out. In total darkness his experimental plants would respire as an animal does, burning sugar to give them calories for work, disposing of the combustion gases into the air. The excess carbon dioxide coming out of the glass chambers was thus a measure of the combustion, a measure of sugar burned. Transeau's notebooks told him how much sugar typical corn plants of varying ages would burn in a day.

It was simple now to work out how much sugar would have been burned by ten thousand plants in one hundred days, and soon Transeau could see a second glistening white pile beside the first, a pile of sugar the plants had first made and then burned. This second pile weighed 2,045 kilograms, so the two piles combined weighed 8,723 kilograms. This was all the sugar made by the cornfield that summer. Now the end was in sight. 8,723 kilograms of the sugar glucose represent 33,000,000 calories, but the man who had measured the sun streaming onto Illinois had found that one acre in a hundred days of summer received 2,043,000,000 calories, more than fifty times as much. If you put one of these figures over the other and multiply by a hundred you get Transeau's result, which was that corn plants, on prime land in Illinois, where they were given every care and attention, were only 1.6 percent efficient.

And so to our amazement we find, not the 20 or 30 percent efficiency of a steam engine, not some super efficiency suggested by ideas of survival of the fittest or the marvelous workings of nature, but a miserable 1.6. Could the scholar in his armchair have got his sums wrong? People have made all Transeau's suggested measurements on real crops, not only corn but other high-yielding plants such as sugar beets, and they have come up with the same general answer: about 2 percent. They also measured the rates of sugar production in photosynthesis more directly, by monitoring the flow of raw materials and waste products to and from the plants, and numerous studies have confirmed the estimates from crops. Our rich productive crops on rich productive soil are only 2 percent efficient.

Perhaps there is something wrong with agriculture. Perhaps it is only plants, grown in unnatural conditions that are so abysmally inefficient. But there is no escape this way either. It is harder to measure the efficiency of wild plants than of crops, but it can be done. You cannot harvest a field of wild plants all of the same age, as you can with corn plants, but it has proved to be not beyond the wit of computer-minded man to make samples and calculate the potential wild crop. We now know that wild plants do about as well as tame plants. A very rough figure of 2 percent describes the efficiency of them all, when they grow in very favorable circumstance. Most wild plants achieve nothing like the 2 percent of agriculture because they do not have it so good. So it is ours to reason why. What curious circumstance prevents 98 percent of the sun's energy getting into the living things staked out to wait for it in such eager array?

What we know of these things has been told us by laboratory people. A plant is grown in a glass chamber, with rigid controls on all the conditions of its life so that it is comfortable and not disturbed; like a baby in an incubator. The breathing of the plant is monitored by measuring the gases it takes and gives to its chamber. When it is busy converting energy by making sugar from carbon dioxide and water, it releases oxygen that sensors can detect; when it is respiring in the dark it releases carbon dioxide. You can do wet chemistry on samples; you can make a plant use a radioisotope of carbon then measure activities; or you can wire the container to the fine expensive electronics of a modern analyst's laboratory, but, whatever way the measurements are taken, one can infer the rate at which the laboratory plant makes the sugar "glucose," and hence the rate at which it fixes energy. Using a water-plant, such as a tiny green alga, makes things easier because the water simplifies the chemistry. Then you shine lights of known intensity into its glass incubator, recording precisely what it does.

A first startling discovery is that half the kinds of light shone on the plant have no apparent effect on its chemistry. Half the total energy of sunlight is in the red end of the spectrum; what we call infrared light. We cannot see this light, but it floods down on us as warm rays, of low intensity it is true, but together adding up to half the energy getting to us from the sun. If red lamps are shone on the plant in its water bath, the chemistry of the water does not change. Plants cannot trap the energy of the far-red wavelengths any more than we can see them. Plants use only "visible" light.

We have obviously found one of the reasons for the inefficiency of plants, but we give a Darwinian biologist a curious question to answer while we are at it. Why should plants be made like people's eyes so that they only make use of "visible" light? Plants must operate according to the rules of our Darwinian game, striving to wrest the largest possible number of calories from their surroundings so that they can turn them into babies. They have been refined by natural selection to do this for a few thousand million years and should be very good at it. And yet they seem incapable of using half the energy pouring down on them. Odd.

When this discovery was first made, an ingenious idea was put forward to explain it. Plants, it was noted, had all first evolved in the sea, and red light does not penetrate very far through water but is rapidly absorbed. Any skin diver knows that everything looks blue down below the surface. A plant growing in an underwater place never has the redder rays shining on it and must do all its work of living with the bluer half of the spectrum. So, it was argued, the ancestors of all plants evolved to be able to use only the energetic rays that penetrate water, essentially the visible light. Plants, however, have now lived on land for several hundred million years, and it is very difficult for a biologist to believe that in all that time they could not adapt to this new brighter world with its red light. Fortunately for our peace of mind, modern physical chemists have come up with a better explanation.

The process of fixing energy (what we call "photosynthesis") involves violent disturbance to electrons as they spin in their orbits round atoms, and it takes a fierce pulse of energy to do this. The radiations of visible light are intense enough to fix energy, but the radiations of the red are not. Life, not for the only time, bows before the harsh reality of physical laws and does what it can with only half of the energy coming from the sun. The red light can warm plants, and does; it also evaporates water from them; helping drive the plants' circulation systems; but that is all.

Since the laws of physics let plants use only half the sunlight, we ought to amend our efficiency calculation accordingly. We double the calculated efficiencies of wild vegetation and crop plants alike; bringing them up from a miserable 2 percent to a nearly as miserable 4 percent. Steam engines and automobiles still manage 20 percent or better, and the greater part of our question about the inefficiency of plants remains.

The next enlightenment to come from laboratory science is that the efficiency of plants depends on the strength of the light. If one shines a very dim light into the laboratory bottles containing the plants, say the light of the dawning or twilight, the plants do amazingly well. If one calculates the efficiency with which they are using this meager resource of light, one may well find that they are doing as well as 20 percent efficient or even more. This does not compare so unfavorably with steam engines and automobiles, particularly when one reflects that a plant has to do its own maintenance as it works, whereas steam engines are made and looked after by others.

So we learn that in dim light the efficiency of plants compares quite favorably with the efficiency of man-made machines. They are not very productive in dim light, of course, because the total energy available is so slight. Twenty percent of very little is still very little, and dim light means poor production of sugar. But plants in dim light yet use what energy there is available to them with tolerable efficiency. Why then do they not maintain this high efficiency when light is abundant and the potential riches in sugar to be won are very large?

If brighter and brighter lights are shone into the plant incubators, the rate of sugar production goes up. This we would expect. But the efficiency progressively falls until it levels off, not at 2 or 4 percent, but at about 8 percent. It is still at about 8 percent when the very highest rates of photosynthesis, of making sugar, are reached. Eight percent of an optimum amount of light gives the highest flow of energy into living things that the bottled plants can be made to achieve. If the plants are given still more light, both their efficiency and the rate of production fall, and a time comes when production ceases altogether. That too fierce a light should stop the plant working completely is not surprising. Presumably the plant is being cooked. It is the low efficiency with which light of optimum brightness is used for which we must find an explanation.

At this stage in the research, our original problem has been compounded rather than solved. We began by asking why crops and vegetation were so inefficient at handling the sunlight with which Providence provided them, and we have not got an answer yet. What we have done is to show that plants are much more efficient at handling dim light than they are at handling the noonday sun and that algal cultures in laboratory incubators may be twice as efficient in bright sunlight as is a field crop (8 percent as opposed to 2 percent or 4 percent depending on the wavelengths supplied). Why are all plants comparatively inefficient in bright light? Why are all plants more efficient in dim light? Why are algal cultures in laboratory incubators twice as efficient as wild vegeta-

tion? The last question is the easiest, and we will take it
first.

An algologist once taunted his collegues, and tempted
the public, with the figures from laboratory experiments
with algae. See! These plants are 8 percent efficient—
far, far better than the corn and the other plants we eat!
It is foolish to grow inefficient crops when we could all
fatten on green algal scum instead! This theme recurs in
newspaper articles about the world food crisis. It is a
myth that is probably as impossible to eradicate as the
myth that *Tyranosaurus rex* was a ferociously active
predator. But myth it is. Algae are not more productive
than other plants.

The catch about the algal culturing is that it is the culturing that leads to higher average efficiencies, not the
algae. Any actively growing plant that can be introduced
to one of the small laboratory cultures will do as well as
the algae. A whole seedling can be put in a small laboratory container, made comfortable, and it will convert the
energy of light to the energy of glucose with an efficiency
of 8 percent or so, depending on the wavelengths supplied. A piece cut out of a leaf can be made to do the
same on its own in the nutrient solution, away from its
parent plant. When conditions are the same, algae are
no more and no less efficient than the crop plants with
which they were so favorably compared.

We now know that any healthy young plant, corn included, which grows in a well-watered field with enough
fertilizer, does as well as the algae (or any other plants) in
the incubators. Its efficiency is that same rough 8 percent of the laboratory cultures. But the special thing
about the plant in the field is that it grows old. When it is
old it feels it age and does not work very well. So the
average efficiency over its lifetime has to be much less
than the 8-percent efficiency of its youth.

At the start of Transeau's hundred days, his Illinois acre was bare of plants and there was no production. At the end of a hundred days there were ten thousand senile individuals who were not doing very much. Somewhere in the intervening time the field was nicely covered with fresh green leaves turning in their 8 percent, but the average for the whole hundred days had to include the beginning and the end, which brought the average efficiency down to 2 percent. Wild vegetation in temperate latitudes faces the same harsh reality: a spring without leaves, an autumn with pretty colors but diminishing green.

The great deception concerning algal culture came very largely from the accidental circumstance that it was convenient for plant physiologists to use fresh-water algae in their experiments. Such cultures are not good ways of producing food (even if we wanted to eat green scum) because culturing requires massive amounts of work and energy compared with conventional crop husbandry. If these inputs of energy were fed into the efficiency equation, we would find that the calculated efficiency was drastically lowered. Algae are no more efficient than any other kind of plant. The answer to our third question is that crops and wild vegetation are less efficient "over-all" than cultures or growing seedlings because of the physical vicissitudes of life, of bare ground in spring, of old age before the winter, of shortage of water and nutrients, of the debilitating presence of neighbors.

Now we must solve the mystery of the dim-light efficiency and the failure of even the favored young to do better than 8 percent. We can find a plausible answer to both of these questions by pondering the supply of raw materials a plant uses in the essential chemistry of photosynthesis. Plants make sugar out of carbon dioxide and water. When water is in short supply the plants grow miserably, as we all know. But when water is abundant it is available to plants in virtually unlimited amount. The other raw material, carbon dioxide, however, is always scarce, even though it is always present. Carbon dioxide is a rare gas. It is present in the atmosphere at an average concentration of about .03 percent by volume, a quite tiny proportion. And carbon dioxide is the essential raw material out of which plants must make sugar. Plant leaves are thin and pierced with multitudes of tiny breathing holes (stomates) for they must suck in carbon dioxide from as many directions as possible if they are to keep their sugar factories going. Even so the rate at which they can soak up the precious gas is strictly limited. It seems reasonable to suggest that it is this shortage of raw material that sets a limit to the sugar-producing powers of plants growing on even the most favorable sites. Plants are inefficient as machines for converting sunlight because they face a shortage of raw materials.

When a plant is grown in dim light, its energy factories cannot work very fast, this being the simple consequence of lack of their light "fuel." In dim light they have carbon dioxide to spare, and only considerations of thermodynamics and plant chemistry inhibit the rate of photosynthesis. The plants in this case turn out to be highly efficient. But as such plants are given more light, their demand on the carbon dioxide supply quickly grows, until they soon are using it as fast as it can be extracted from the air. At this moment plants are working as fast as their factories can be made to run. They are then about 8 percent efficient. If they are given more fuel still, as by shining the noonday sun on them, they can only waste the surplus, degrading it to heat, pouring it away.

We can test our hypothesis that carbon dioxide limits the productivity of plants by pumping a little extra into our plant incubators and seeing what happens. If we do this, the rate of sugar-production goes up and the efficiency of energy conversion in bright light is slightly increased. If we give the plants too much carbon dioxide, we suffocate them; but this need not disturb us. Plants have evolved in a world in which carbon dioxide is scarce, and their chemistry has adapted accordingly. Yet the dependence of sugar production on the carbon dioxide supply is clearly shown by these experiments.

It is well to insert a small word of caution about the generality of this result. The logic that so scarce a commodity as carbon dioxide ought to limit the rate of production is sound, and the experimental data are convincing demonstrations that we are on the right lines. But some of the consequences of a shortage of carbon dioxide are very complex and may impose second-order restrictions on photosynthesis. Plants must "pump" large volumes of gas as they extract their carbon, and this pumping may introduce its own restraints. Flooding the plant tissues with oxygen in the flux of air will have its own consequences for reactions dependent on chemical oxidations and reductions. Opening the stomates must result in the escape of water. And so on. All operations that boost production in the plant factories must involve their own constraints, and we can expect many fresh limits to appear as plants evolve to make the most of the carbon dioxide supply in different circumstances. These possibilities are reflected in many modern debates about alternative "pathways" of chemical synthesis in plants. But with this bit of mealy-mouthing we can yet say that plants are generally inefficient as converters of energy because carbon dioxide is a rare gas in the terrestrial atmosphere.

This finding is of great significance to practical people for it means that there is a very narrowly defined limit to the possibilities for growing human foodstuffs. Our ultimate yields are set by the carbon dioxide in our air, and there is nothing we can do to push plants to do better. Our so-called high yielding strains of wheat and the rest are in fact no more efficient than the wild plants they replace, whatever the gentlemen of the green revolution may claim. All that the agriculturalists have done is to make plants put more of their total capital of sugar into parts that people like to eat. A high-yielding wheat makes more grain at the expense of stalk, roots, and the energies to defend itself against pests and weeds. The finest efforts of science have not made any plant one jot more efficient than those nature made.

To a biologist brooding on the great conundrums of life, the inefficiencies of plants have a different message. The fuel supply for all life is restricted to some small fraction of what comes from the sun. A theoretical upper limit is about 8 percent, but this will be reached only for very short periods in very small places. All plants face youth and senescence, and virtually all face the changing seasons. All suffer at times from shortage of water or nutrients; none works at full efficiency for long. When we think of the average condition of life on earth, we think of deserts, mountainsides, and polar ice caps, as well as fertile flood plains. The average productivity of the earth must be very low, certainly much lower than that 1.6 percent of Transeau's cornfield. Probably only some fraction of 1 percent of the solar energy striking the earth actually gets into living things as fuel for plants and food for animals.

When we try to explain the numbers and kinds of plants and animals, we must remember this great restriction in the fuel supply. Plant-eating animals, for instance, can get only a small portion of the sugar made by the plants on which they feed. This is hard to measure, but practical people generally accept an upper estimate of about 10 percent. We may think, therefore, that on good pasture land, herbivores get 10 percent of 2 percent of the sun. A tiger hunting those herbivores might in theory get 10 percent of 10 percent of 2 percent of the sun. And so on up the food chains.

We come then to the proposition that the numbers of the different kinds of plants and animals on earth are set by the amount of carbon dioxide in our air. Carbon dioxide sets the rate of plant production and is hence the ultimate arbiter of the food supply of all animals. If our earth had been forged with more carbon dioxide at its surface, the plants would have delivered more food and the opportunities for animals would have been greater. We might even have had tiger-hunting dragons then, and the ferocious tyranosaur would have been less mythical. But the chemistry of the earth's surface keeps the concentration of carbon dioxide low, by mechanisms quite out of reach of plants and animals. And so the answer to many general questions about the numbers of animals as well as to the inefficiency of plants becomes, "Because there is very little carbon dioxide in our air."

# Chapter Five. The Nation States of Trees

THE traveling naturalists of the eighteenth and nineteenth centuries brought back to their European universities the astonishing news that the plants of the earth were organized by continents and geography into something like nation states. Over enormous blocks of territory there spread formations of plants whose members were all of the same shape. This was, on the face of it, an uncalled-for thing. Naturalists knew, of course, that plants had a number of clearly recognizable shapes, that there were conical trees and round-topped trees, straggling herbs and bunched-up herbs, but that these should be organized, as it were, into nation states of plants was a new idea. It gave rise to some intriguing thoughts about the qualities of life on earth, the echoes of which are with us still.

A naturalist slowly making his way round the world in a sailing ship would typically have come from an agricultural countryside, whose wilder parts held deciduous forest, the familiar woodlands of North Atlantic man that apparently represented the wild forests that once covered all his agricultural land. These woodlands were dominated by fine, spreading trees, whose branches formed a canopy overhead in summer but lost their leaves in winter. Underneath were scattered bushes, a few creepers, and a forest floor tinted brown by fallen leaves but blessed every spring with a multitude of flowers that filled the open spaces under the bursting leaves with fragrance. These woodlands were familiar, describable things. The names of the flowers and the trees might change from place to place, but the shape and feel of the woods was the same over much of Europe. But when our traveler sailed to the equatorial regions, he would find the plant array strangely different.

The trees of equatorial lowlands are vast by the standards of the European woods. As a young man I used to think the great horse-chestnut that stands in front of the chapel of King's College in Cambridge was a noble tree, but then I wandered in the rain forests of Nigeria and there came an inevitable day when I realized that that splendid chestnut would fit, all of it, under one of the lower branches in the Nigerian forest. Yet the European travelers saw even stranger things than daunting size. The equatorial forests are evergreen. Under the tallest trees are shorter trees, and, under these, shorter trees still, though still tall enough. The interlocking leaves shut out the sun, letting through only a dim green light so that the African traveler Stanley felt as if he was in a great cathedral, the rising Gothic columns of the trunks around him being lighted through windows of green stained glass. Orchids and bromeliads clutch at the trunks of trees with mossy roots. Lianas hang above as unaccustomed nets, and the air is filled with alien sounds.

When the traveler went northward, to Scandinavia, Russia, or Canada, he found whole nation states of forests given over to the cone, to the dark green brooding ranks of needle-leaved conifers. Further north he came out upon the treeless regions often with astonishing abruptness. Around him then stretched the open plains of the tundra, sometimes gay with flowers, sometimes even rich with berries, but always a land from which trees were banished.

There were prairies to visit too, open plains of long, waving grasses stretching from the Missouri River to the Rocky Mountains or all across Argentina to the Andes. There were African savannahs as large as European kingdoms where dry grasses rustled between the flat acacia trees. And there were bushlands like the maquis of the Mediterranean, and the chaparral of California, or some of the heaths of Australia whose bizarre species yet conformed to the shape of similar bushes in southern France.

Each of these great formations of plants seemed vividly distinct, holding on to the traveler's memory— foreign nations of plants, each occupying its separate region of the earth. One could actually map the plant formations. Much of the task would be easy: there was tropical rain forest from the mouth of the Amazon to the mountains of Peru; there was prairie from roughly the line of the Mississippi to the mountains in the West; the deciduous forest line of Europe ran from the corner of the Black Sea inland and westward to the Pyrenees. The map might follow the natural contours. Or the plants themselves may set the co-ordinates, as happened with the arctic treeline where the edge of the forest was so notable that many people had left written records of it. When data were too scant, as in southern Asia where reports were confused, lines would be drawn as well as possible.

The map of the world that resulted showed all the plants of the earth marshaled in great divisions, that came to be called "formations." They were drawn up like nation states, the individuals in each conforming to some common plan as if put there by a master hand. As long as we accepted that this was the work of a Creator, this was perhaps well enough; he presumably knew what he was doing. But if all those individual species of plants were fashioned by natural selection acting on random diversity we had some problems to explain. How could there be different formations staring at each other across frontiers like nations at war?

While Darwin was writing *On the Origin of Species*, a French botanist was thinking of these things, as a sideline to the great work of his life. He was Alphonse de Candolle, a taxonomist, an herbarian botanist who traveled little but who had at his disposal the great Paris collections representing all the known plants of the world. With this material he was making what in fact turned out to be the last attempt by one man to describe all the plant species known to science, eventually published as his master-work *Prodromus*. As Candolle described the species, he became acutely aware of the curious way they were regimented into formations, and he knew that this was something to be explained. Some aspect of the weather seemed the obvious answer. Shortage of water, an accident of geography, should certainly explain the desert formations and, perhaps, grasslands, but something else was needed to explain the rest.

Candolle thought the answer must lie in that other easy-to-measure parameter of weather: temperature. He went so far as to say that there must be critical changes in the heat regimen at particular times of the year in the frontier lands that accounted for the changes between one formation and another, and he even guessed at the isotherms that would be found to follow the formation boundaries. The watch on the world's weather had hardly started in the eighteen fifties, giving Candolle little chance of drawing realistic temperature maps. But we have used modern records to plot Candolle's isotherms in my laboratory, which result in a map closely similar to a primitive vegetation map of the world.

Candolle's writing on the subject stimulated the thoughts of pioneer meteorologists who were then just beginning the watch on the world's weather. As with all sciences, one of their first tasks was to classify the phenomena of their discipline, and classifying weather must surely mean to map it. But how does one map weather—without stations in every country, without satellites in orbit, when the fronts of weather seem in endless flux? The answer the meteorologists chose was to accept Candolle's conclusion that it was weather that set the boundaries to the nations of plants. They mapped the plants and called them "weather."

Most notable of the climatologists of this period was Vladimir Köppen of Vienna, who not only acknowledged Candolle's conclusions and used his maps but also agreed that the five main kinds of formation recognized by botanists must represent five principal classes of weather. Tropical rain forest, hot deserts, temperate deciduous forest, the boreal forest of Christmas trees, and the tundra became climatic types A, B, C, D, and E. All the lesser formations such as the bushlands of the maquis and the chaparral represented subdivisions of one of the main climatic types. All maps of climate, even those used today, reflect these original decisions of Vladimir Köppen and his peers.

Any contemporary atlas will contain, next to the map of the world's vegetation, a map of the world's climate. And there will be the shadows of Candolle and Vladimir Köppen, patterned across the pages in blocks of color. The rain forest, the tundra, and the regions given over to the cone will be on the one map; the climates A, B, C, D, and E on the other. The two will match because they are the same map.

A turn of the page will reveal the same map for a third time, but now it is called a map of soils. As one goes from one part of the world to another, the soil under one's feet changes strikingly. This is not just a matter of subtle chemistry of rock; nor can it be seen only by those who know of tillage, drainage, and manure. The soil of one place is as obviously different from another as a pine tree is different from an oak.

Dig a pit in the wet tropics and the sides show red, dig it in middle Europe and they are brown. In the vast conifer forests of southern Russia or Canada the sides of the hole are striped with extraordinary colors. There are fragrant needles and black humus at the top of the ground in the boreal forest but then a broad band of glinting white or gray perhaps six inches thick before one comes to depths of brown and rust. Russian peasants have long coped with this soil and know its hostile properties for farming. They called it the ash soil (*podzol*) because when they ploughed it the ground looked as if it was covered with ashes. It takes no scientific knowledge to see that this oddly striped ground is very different from the brown and red earths of more southern climes. North of the treeline, the striped ground is left behind and beneath one's probing spade is found a ringing block of ice covered by a thinnish mat of sodden humus, a tundra soil.

Soil people, like weather people, needed to map these striking oddities as a first step toward an understanding of them. But setting out to map a soil is nearly as daunting as mapping weather. It is true that soils do not move about in the uncooperative way of weather, but they are covered with debris and plants and cannot be seen except by digging a hole. And digging takes time. So one digs as few holes as possible and uses the general appearance of the land and its plants to guide one's mapping between one hole and the next.

I have mapped soils for a living myself, in the lovely wilderness of northern New Brunswick. We used air photographs to guide us, picking out the lines where one subtly different kind of forest met another by staring at matched pairs of photographs through a stereoscopic viewer. Then we dug our holes where the photographs of the forest told us to, and marked on the photographs the spread of the soil type to the line showed us by the trees. We mapped the plants and called them soil.

In the New Brunswick forest we were looking at small details in the different soils, and some of these details were so subtle that they could be discovered only after long observation. All our soils were podzols, and all were under coniferous trees. We were distinguishing between different kinds of podzol, but those who made broad soil maps of the whole world worked in the same way. They mapped the plants and called them soil. Thus it is no surprise to find the soil map of the world in the family atlas looking remarkably like the map of vegetation. And both of these will look like the map of climate because all three are the same map.

Meteorologists and soils people have got away with their grand mapping deception for so long because their maps work. If one goes to the middle of one of those colored blobs in the atlas one will find that the local soils and climate are, broadly speaking, what the legend in the atlas says they will be. It follows, therefore, that the assumptions under which the maps were made are good assumptions. The weather is the arbiter of soils and of those nation states of trees.

It is easy to accept that the weather controls our vegetation. Indeed, farmers were well ahead of scientists in this. But it is nearly as easy to see that weather and plants control the soil. The strange podzols, for instance, are found only in places with cold winters, mild summers, and vegetation of the cone. The ground in these places is always covered with fallen needles, which rot very slowly, yielding acids to the soil water that make the whole top section of the ground literally acid. Earthworms and their like cannot live in this acid ground, and thus there is nothing to churn it up. The earth remains, mineral flake on mineral flake, with little movement, year after year. The cold, acid water percolates through, washing and washing, a subtle chemistry that bleaches the top inches to the color of ash. But where it is warmer, and where there are no conifers to smother the ground with slow-rotting needles, then the soil is not so acid and earthworms can live and churn, then we get the brown soil familiar to most western farmers. And in the tropics the warm sousing rains work a more potent chemistry still, as they have done since time immemorial, year round. This chemistry leaves behind only clays stained with insoluble red oxides of iron and aluminium; and we see the red mud familiar to those who have traveled where there is heat.

Our earth, then, is covered with a patchwork quilt of weather, and the patterns of this patchwork are faithfully reflected by the plants and soil. This seems a safe enough statement, but it leaves some mysteries to be solved. Why, if weather sets the pattern, do very clear changes occur from one place to the next? The differences between a spruce forest and one of oaks are quite extraordinary, and the differences between either and a treeless tundra are odder still. Why such mighty changes in form between places? Why do virtually all the plants of each place respect the prevailing fashion even though their evolutionary ancestry may be different? The fashion can change from one place to the next with astonishing speed. In an hour or so of driving north across Connecticut and into southern Canada the shapes in the flashing blur past the car windows will change from broad-leaved forests like those of Old England to the seried ranks of conifers like those of Sweden. How can there be such abrupt changes? Why do the plants of the earth gang up together in places of their favorite weather? Have they been organized by some unseen hand, some guiding principle of life that regiments them like the peoples of the nation states that the formations mimic? Since the first botanist made a plant map of the world we have pondered this mystery.

A high privilege of living as a scientist in the last decades has been to see great questions such as these suddenly yield. Take, for instance, the matter of fashions in plant shape. Why should trees keep their leaves year-round in the tropics and in cold Canada and Sweden, but the forests in between lose theirs in winter? Why do desert plants grow like organ pipes? Why in the arctic are there no trees at all? The weather is obviously the designer of these fashions, but how can weather cause such curious designs and conformity? Botanists tried for a hundred years without convincing general answers. But now we know.

Consider first the matter of the missing arctic trees. It was never enough to say, "Oh, it is just too harsh for trees there," because succulent meadows of fragile flowers flourish in the arctic, delicate things that could not survive harsh treatment. Someone using words such as *harsh* is likely to be asked what he means by it, and trying to translate the meaning of *harsh* into the experiences of a tree for many years failed to help us understand why trees could not live in the arctic. Perhaps it was too windy; but there are places in that treeless expanse where it is no windier than other areas further south. Nor could one claim extremely cold winters, for there are treeless places such as the Pribilof Islands of the Bering Sea with relatively mild winters, and there are places with forests that suffer winters of thirty below. Permanently frozen ground does not inhibit trees, since forests are growing over permafrost in the valley of the Yukon. Saying that the arctic is too harsh for trees does not get us very far.

Plant physiologists have suggested that part of the answer may lie in the short growing season, in the very limited opportunities afforded a plant to make sugar with the energy of the sun. Arctic plants exist on very short commons, which means that they must be economical in spending their food calories. A tree has a large body to maintain, an unproductive, parasitic body of woody trunk and limbs. This may be just too expensive to be afforded by the income of calories that can be won in an arctic summer. No doubt this is part of the explanation, but it is not completely convincing. There are woody shrubs in the arctic such as the prostrate willows, which may have a very considerable mass of unproductive woody stem snaking flat along the ground. If a horizontal trunk can be maintained, why not a vertical one?

The convincing answer comes from the work of David Gates, a physicist who turned coat to become curator of a botanical garden. Gates thinks of plants, and animals too, as bits of mechanism that must balance their heat budgets. A plant by day is staked out under the sun with no way of hiding or sheltering itself. And all day long it absorbs heat. If it could not lose heat as fast as it gained it, then it would heat up, be cooked, and die. Plants get rid of their heat by warming the air around them, by evaporating water, and by radiating to the atmosphere and the cold black body of space. The plant adopts a working temperature that is a function of the rate at which it absorbs heat and the rate at which it gets rid of it again. It is absolutely vital that the plant be able thus to balance its heat budget at a temperature tolerable for the processes of life.

A plant in the arctic tundra lies very close to the ground in the thin layer of still air that clings there; but a foot or two above it are the winds of arctic cold. All the tundra plants together are absorbing heat from the sun, tending to warm up, probably balancing most of their heat budget by radiating to space, but also warming that thin layer of still air that is trapped among them. While they are close to the ground, they can balance their heat budgets. But if they stretched up as a tree does, they would lift their working parts, their leaves, into the streaming arctic winds. Then it is very likely that they could not absorb enough heat from the sun to avoid being cooled below a critical temperature. Your heat budget does not balance if you stand tall in the arctic. That is Gates's explanation for the absence of trees there. It is almost certainly correct.

Gates's thinking of heat budgets also helps explain some of the other characteristic shapes of plant formations. A desert plant faces the opposite problem to an arctic plant, being in danger of overheating. It is short of water and so cannot cool itself by evaporation. One of the solutions to this problem is the familiar tall, stick-like shape, a rod pointing at the sun. This shape gives the plant the smallest possible surface exposed to the incoming radiations but the largest possible surface away from the sun from which it can radiate heat out.

Rain forest trees do not have to worry about the sun, because they have ample water for their cooling systems and can keep fragile flat leaves facing the sun year round. In temperate latitudes, the same flat leaves work well in summer, when conditions may well be almost tropical, but the design encounters grave difficulties in cold weather. A broad, flat leaf presents a large surface to the black body of space so that the radiation loss is likely to be high, but cooling by convection is a more serious problem still. Gates demonstrated this by placing casts of leaves in wind tunnels.

Gates's casts of leaves were of silver, which let him

measure heat loss directly as he exposed them to winds of different speeds and temperature. A broad, flat leaf proved to be a poor thing to put out into cold winds, causing a turbulent flow that extracted heat from the leaf quickly. This suggests that broad leaves cannot be kept at a suitable operating temperature in times of wintry sun. It is apparently better for a tree to withdraw as many calories as possible from the leaves in the fall, drop the empty husks, and grow a new set of pristine energy transducers in the spring.

Further north still, the coolness and shortness of the summer growing season makes this strategy less economically successful. The answer then is the round needle-shaped leaf borne in thick clusters, the needles of the coniferous evergreens. Casts of these needles in wind tunnels lose heat less rapidly than casts of flat leaves, a consequence of a smoother flow of air. Moreover, being round, these needles radiate heat in all directions, to the comparatively warm surfaces of their neighbors and to the ground as well as to the cold of outer space. A spruce tree thus loses heat more slowly than an oak tree. It can keep its needles all winter, ready to take advantage of brief, warm sunny spells. During these spells it can maintain a temperature high enough for its energy transducers to produce a worthwhile income of sugar calories. Even though a spruce tree may not be so good a worker in fine summer weather as an oak, its strategy probably gives it a better overall return when spread over the year at high latitudes.

So now we have a working model to connect the shapes of the great formations of plants with temperature of the air, availability of water, and the presence or absence of seasonal differences. We know how climate exercises its control on plant shape; the work of Candolle,

the botanist, and Köppen, the climatologist, is fully vin-
dicated. We know what sets the fashions in plant design.

And so back to the frontiers of the nation states of
plants. Why should a fashion in plant shape hold sway for
continental distances, and then change so swiftly to
another fashion, as our maps show? Because the weather
changes. Yes, but the weather is a fluid thing which
wobbles its way across whole continents from one season
to the next. So how can the plant formations have these
remarkable edges? Common sense expects a diffuse
blending of any pattern set by the weather. But we get
nation states of trees; can there be some hidden organizer at work after all?

Some of the more dramatic implications of a map of
vegetation of the whole world shrivel with a little cool
thinking about how the map was made. The phenomenon is one of simplicity of large scale. In drawing the
grand vegetation map, we took advantage of natural features such as mountain ranges and ocean coasts to guide
our hands in drawing boundaries, a procedure that let us
ignore the subtle changes in vegetation that were taking
place as these natural features were approached. Where
we had no geographic boundary to follow, we drew the
lines as best we could, which is to say we chose arbitrarily. Most of those nation-state boundaries between the
formations are merely map-makers' conventions, boundaries drawn at convenient places across regions where
the rates of change in the blending plant communities
are particularly rapid. The boundary on the map represents a diffuse blending of the formation of plants on the
ground.

But what of the arctic treeline and of that remarkably
rapid transition from decidious forest to Christmas trees
in New England and Canada? Here in the open conti-

nental spaces there are no geographical boundaries that the different trees and tundra can follow. Here the weather marches to and fro with the seasons, and yet the forests have edges we can actually see. How do the plants achieve this?

A treeline may sound an abrupt line but over a continental expanse such as Canada or Siberia, the forest shrinks in stature for many miles approaching its edge—the trees becoming smaller and smaller, spread thinner and thinner. There is always much interfingering between the forest and the tundra as stunted trees follow the valleys of streams whose dendritic patterns stretch far into the otherwise treeless plainsland. Yet on a scale of miles, the treeline is a pretty definite boundary. North of this line a tree apparently cannot balance its heat budget, but south of the line a small tree can. Is it suddenly "nicer" where the line of trees is such that one jumps from the bleak arctic and lands in a balmier place among the trees? How do the trees know where to stop? We know the answers to these intriguing questions too, many of them from a recent piece of work by the Wisconsin climatologist, Reid Bryson.

Bryson made a map of the climate over the arctic, not as the pioneer climatologists had done by mapping plants and calling them weather, but by ignoring the plants and mapping the climate directly. He measured air temperatures, wind speeds, and air mass trajectories; and he did it from a network of stations in many parts of the American arctic for ten continuous years. Bryson's computer generated, from these data, a map of the average position of the front of the arctic air mass in summer. This front, of course, wandered widely from year to year and from season to season, as fronts do, but its average summer position, as drawn by the computer, followed with astonishing accuracy the line of the trees that botanists had mapped before. The synchrony between air mass and vegetation was directly demonstrated. But Bryson's map did even more than this, because it let a reasonable man understand why the treeline was where it is. Trees live long, and the forest moves only with the slow tempo of their generations. We must imagine that the treeline is in flux with this tempo, too slowly to be noticed in the span of human lives. It is the average position of the rapidly moving front that is of importance, for on the average a tree that is too far north will suffer calamity and on the average a tree that is far enough south will survive. So the forest recognizes a sharp edge in the weather, for all the wandering of the fronts in secular time.

Bryson's air mass map also explains why there is so neat an edge to the south of Canadian Christmas trees. The swelling arctic air usually reaches this point in winter. In summer, the arctic air shrinks back toward the pole, and the edge of its front sets the northern limit of trees. But in winter the arctic air pours southward, bringing its icy presence to the Canadian border lands. The needle-leaved trees endure this, shedding snow, taking what warmth the wintry sun allows, patiently ready to go into full production when the arctic air retreats on the first day of Canadian spring. But the strategy of the broad-leaved deciduous tree is not suited to this inundation or to the short frost-free period that is all that is allowed for summer. So the southern edge of arctic air in winter is another diffuse and floating edge whose average position is read by the trees. They do not read this edge so well as they read the more arctic one, however, perhaps because it is a winter edge, present only when the trees are dormant. During the growing season in the lake states, for instance, the arctic front lies far to the north, ebbing and flowing across the treeline.

The boreal and deciduous forests of middle America blend together across some hundreds of miles. But botanists do recognize a transition zone and the map-makers have drawn their line for this roughly where Bryson plots the winter front of the expanded arctic air. To the north of this line, on the average, the strategy of needle-leaved evergreen pays off best; to the south, on the average, the strategy of having broad, efficient leaves that are lost in winter is best. And so there is coniferous forest to the north of the line, deciduous forest to the south of it, and a hundred miles of transition between.

Thus the organizer of the nation states of trees is seen to be the weather. The plants living in the same climate all adopt a common shape over evolutionary time because this shape gives the best trade-off between the needs of efficiency of production, the use of the water supply, and the balancing of heat budgets. Where the climatic boundaries are distinct like the edge of the arctic air, then there will be distinct borders to plant formations. This will also happen when the earth process throws up a barrier to plants and climate alike; a mountain or a sea coast. But when there is no natural boundary, and when climate does not march with such annual precision as in the arctic, then there will be a grand blending of vegetation and the maker of maps must draw an arbitrary line to complete the neatly patterned pictures in your atlas.

There are no great mysteries behind the nation states of trees anymore. The only mastermind is the weather. And the agent who establishes most of the frontiers is called a cartographer.

# Chapter Six. The Social Lives of Plants

HARD upon the discovery of the nation states of trees came the beginnings of plant sociology.

Plants grow in communities. They grow up together in complex patterns. All plants have neighbors, close neighbors with whom there must be some accommodation or none would survive. Each kind apparently raises its young in the presence of the others, or at least the plants grow up together and share the living space. The lives of plants in communities do not seem to be chance affairs, for we can see the same patterns of life repeating themselves over and over again. It is almost as if we are looking at the workings of a social process: the social lives of plants.

Any forest of a temperate land is made up mostly of just one or two kinds of trees that are overwhelmingly predominant; we can justly say that the forest is dominated by them. Other kinds of trees exist there, but they are likely to be scarce. There is also likely to be a host of other kinds of plants, of creepers, and bushes, and spring flowers. Is not such a forest a society of plants dominated by those one or two kinds of aristocratic trees?

Any oak wood in southern England has much in common with other English oak woods: they are all dominated by oaks; they all have familiar bushes under the trees such as hornbeams, they are even likely to have the same kinds of spring flowers. One has an impression of a social organization dominated by oak trees. Similarly, a spruce forest in Norway is very like another spruce forest in a different part of Norway; here is a social organization of species dominated by spruce trees. The woodlots left on good farmland in the American mid-West are likely to be dominated by beeches and sugar maples, though a total of eighteen other kinds of tree can be found scattered through a big patch of this rich American forest, and as many as ninety-five species of shrubs and herbs have been described in the society of the beeches and maples.

Here is organization repeating itself. Botanists used the word "association" to describe the exciting idea that had come to them. There was an oak association in England, a spruce association in Norway, a beech-maple association in the mid-West, and of course many others. Patch after patch of natural forest conformed to a general description of the association type; the same dominance, a similiar physical structure, a significant proportion of a common pool of species of lesser plants. The mind slides easily from this useful abstraction, in which associations of plants often live together, to ideas of compulsion. What makes the plants live together in predictable array? Who is the policeman who orders these societies?

That plants in nature live grouped in familiar assemblies is a discovery, an observation of actual fact. But that these assemblies represent societies, put together according to some prevailing social law of life, has the more thrilling status of an idea. It sounds almost too exciting to be true; social lives in plants! Yet there the plant associations are, familiar, describable communities; there has to be some explanation for this.

More than fifty years ago many botanists set out with high hopes on what was to prove the long and difficult quest for an answer to this riddle of plant organizations. They sensed that they were on the threshold of a new adventure of the spirit. They would classify natural order as represented by the plant associations and find its guiding principles. Zoologists and botanists a century earlier had classified the simpler natural order revealed in the different species of plants and animals and from those labors had come the theory of evolution. Might not some new splendid theory come out of the study of whole natural communities?

So botanists set out to master the laws that governed organizations more complex than mere species. Like Darwin approaching his "mystery of mysteries," they felt themselves very near the presence of a great truth. They called themselves "phytosociologists," sociologists of plants.

The first sociologists of plants made their studies in the south of France and in the neighboring Alps of Switzerland and were known as the "Zurich-Montpelier School." Their method was to find a "good" bit of vegetation, a subjective choice that was probably easy enough in a place that had been farmed for five thousand years and where a patch of woods might even have a convenient fence round it. Then they would explore and describe the community, noting which plants were dominant, which were common and which were rare, which were sociable in that they lived in clumps and which were unsociable in that they lived alone. Then they would seek another area of vegetation that looked like the first and describe that too. Then another and another. Soon they had pages and pages of descriptive lists, all of plant communities that looked alike. Then they compared their lists. Each list contained a few maverick plants peculiar to that list; these were eliminated. What was left, dominant and faithful species common to every list, together with a larger array of less faithful but very frequent species, made up the formal definition of a plant association.

The Zurich-Montpelier student had described what he thought of as a sociologist's *species*, taking his association to be as valid a unit as the species described by a museum taxonomist. After all, the museum scientist did no more than describe specimens in a subjective way. He looked at a large number of individuals, worked out an average description, and called the result a species. This is what the Zurich-Montpelier sociologist had done, except that each of his individuals was a whole society.

The Zurich-Montpelier people went on to describe all the different kinds of plant communities in the south of France and in Switzerland in this way. Then they spread further afield, describing each obviously new community as they came upon it, adding to their collection of "associations" as they went, just as museum collectors added to the known list of species. Not only forests were collected but all kinds of communities; moors, pastures, bushlands, tundras, deserts. All places held communities of plants; "associations" could be described in them all. Then came the work of arranging that great collection of "association" species to reveal the underlying natural order. When museum taxonomists had done this with *their* species, they had pointed the way to the theory of evolution. What mighty knowledge might lie behind this new systematics of ecology?

One thing the classifiers of associations could easily do; they could assign associations to one or another of the great formations of which the plant geographers spoke. Associations dominated by oak trees, or beeches and maples, or chestnuts and walnuts, or birches and poplars, would all belong to the formation of temperate deciduous trees. Associations dominated by spruce trees, or pine trees, or larches would go to the boreal forest, the formation of Christmas trees. And there would be associations of the alpine meadows that could be assigned to the true tundra, for all that they were living hundreds of miles south of where most of the associations of the tundra were located. So far the classifiers could go, but no further. The associations could not be fitted together into genera, families, and orders as could museum species, or at least not in ways clear enough for everybody to agree to them. Many tried but none succeeded. It looked as if a new understanding of nature based on plant sociology was not to be. But, meanwhile, another school of phytosociologists went to work as if it was clearly the method that was wrong and not the fundamental idea.

The Uppsala School worked from Sweden across Scandinavia and parts of northern Europe, an area very different from the smiling diversity of southern France and the Alps. This was a land of broad expanses of the brooding ranks of Christmas trees, simple, unlayered, dark forests; it was also a land of wild moors and open tundra. It comprised more wilderness than farm. In such a country it was not so easy to select a choice bit of vegetation to describe because of the land's everlasting sameness with only subtle changes occurring over long distances. So the Uppsala people did not try. Instead they took up the mathematical techniques of random census. Starting at arbitrarily chosen places in the forest or moor, they began to count plants. As they spread further and further in their counting, they found new kinds of plants to add to their lists. But, after a while, they would stop finding new kinds because they had spread over a large enough plot of ground to include almost everything living in the local community. Then they stopped counting. Now they had a list like the Zurich-Montpelier lists, with many plant names, some common, some rare, some sociable, some loners, some dominant and big, some servile and obscure. But this list had been compiled with the impersonal methods of the statistician; it had not the taint of subjective science with which those who chose their bits of vegetation could be charged.

Then the sociologists of Uppsala began comparing many lists, just as those others had done in the south of France. They too identified "species," which they called "sociations" (to avoid confusion with "associations" of Zurich-Montpelier). Then they tried to classify their "sociations" to reveal that Holy Grail of a new truth in science, which had escaped the subjective workers of the south. But they had no more luck. The plant societies refused to be classified in any meaningful way.

Even as they tried to classify, a prosaic truth made itself felt in the minds of the scientists who wrestled with the oddities of plant societies. Their communities, whatever social name they gave them, in fact only reflected physical changes on the face of the land. In a wet bottom one kind of association occurred, on a dry ridge another, on a fertile flood-plain a third, and so on. It was habitats scientists were describing with their elaborate lists, not societies. One particular kind of soil and site was suited to a select list of locally available plants and these would be found growing together wherever those physical conditions for life were repeated. The policemen regulating the plant societies were only the cold facts of physical existence that all living things must recognize. How boring. The debate, however, was kept alive by the long and passionate affair of the vegetation of mountainsides.

It was an explorer of the American West, C. Hart Merriam, who gave us the mountainside problem. Merriam was charged with a biological survey of a part of Arizona in 1889, in the pioneer days when the region was still biologically unknown. In one season, the young Merriam brought back twenty mammals new to science. But he was particularly impressed with the vegetation belts he encountered on the ascent of the 13,000-foot San Francisco Mountain. The mountain has its roots in the Sonoran Desert, a landscape studded with cacti, including the organ-pipe "saguaro," and with leathery or thorny shrubs. But Merriam climbed out of this burning desert through the forest of dwarf oaks on the lower flanks of the mountain until, at 6,000 feet, he came into woods of tall pines, with streams of water laughing through them and a medicinal smell from the carpet of needles underfoot. Six thousand feet; that was all it took for so startling a change from desert to piny paradise. Merriam climbed on, above the pines to a forest of Douglas firs, then still further up until the trees gave out and he found himself on the alpine tundra of the summit. He marveled at these splendid transformations and wrote descriptions of each of those lovely bits of vegetation: the desert, the oak scrub, the pine forest, the Douglas firs, the tundra. He called them life-zones. He might just as well have called them "associations," but the word had not yet come into fashion.

Merriam wrote of these life-zones as if each one was distinct. One might expect this approach if each reflected a special social organization of plants, but it does not make sense if plants are mere passive followers of physical habitats. The physical habitat should change gradually as one goes up a mountain, not in discrete jumps every few thousand feet. The air becomes gradually colder as one climbs a mountain, gradually more windy or rainy. In the Darwinian view of life, individual plant species should find their own levels on the side of a mountain so that species recombine endlessly with the ascent, resulting in the continual blending and merger of vegetation types. But Merriam said there were separate life-zones stacked one upon another.

If "seeing is believing" then many a mountaineer will say that Merriam was right. Belts of vegetation do appear to be stacked upon each other on a mountain if one views it from across a valley in Arizona as Merriam did, on a Galapagos Island, in the Alps, on Mount Cameroon or Kilimanjaro, in New England and Appalachia. Looking across at the Appalachian Mountains in the fall of the year, one sees bands of color stacked on top of each other, striking colors such as the reds of oaks and maples, the dark greens of mountain conifers, the fresher greens of stands of trees whose leaves have not yet turned. The eye can trace these bands of color along the mountain range, band stacked on band, life-zone on life-zone, association on association.

Although it turned out to be pressingly hard to classify the associations of the flat terrain, and although every association closely examined proved merely to be the reflection of a physical habitat, of a piece of ground, a sociologist of plants could take heart from this knowledge of the mountaineers. The mountainside was a place of continual and essentially gradual physical change, but across this physical gradient the plant associations strode. Each belt stretched over a broad swath of mountainside, a plant society held intact for its members until the borders of the domain of the next society was reached. It was the nation states of trees all over again, territorial boundaries and all.

But alas it really was not. Both the mountaineers and putative sociologists of plants were victims of an optical illusion.

Robert Whittaker, now of Cornell, took the trouble to look very closely at mountainsides. He ignored all presumptions about zones and plant societies, plunging straight into the forests of the flanks and making a running census of the principal plants as he ascended. It was a simple and obvious thing to do. Whittaker's running censuses took him right across those apparent zone boundaries so that he should have been able to identify them as discontinuities in his results. But the zone boundaries could not be found. Instead the data showed clearly that individual species of plants came and went with gentle gradualness as one ascended a mountain, that there was that endless blending of species that should result if each kind of plant did its own free thing, without benefit of social organization.

What then of the vegetation belts that layman and botanist alike can see with their naked eyes? There is no conflict. A person looking across a valley at the mountain in the distance picks out what is notable—the level where red oaks are abundant, or pine trees, or rhododendron bushes. The places where these concentrations of color or texture merge are put to the back of your consciousness. You are looking at a spectrum of changing plants as you sometimes look at a spectrum of changing colors in a rainbow. We talk about the violet or the green bands in the light spectrum of a rainbow, but there is an endless progression of wavelengths of light from one end of the spectrum to another. Distinct bands of color in a rainbow are an optical illusion, a convenience for memory and expression. The same is true about the belts of vegetation on a mountain; they do not exist as discrete zones of vegetation.

There is this which is still to be said about the idea of the *association*: some plants do dominate patches of vegetation, and other plants have evolved in the face of this reality. There are species that are faithful to the presence of others because they get their living in association with these others. It is possible to occupy the niche of living-where-the-oaks-grow  or  getting-a-livelihood-in-the-shade-of-the-beeches-and-maples.  To  the  extent that new niches are evolved to take advantage of the presence of other species, there can be associations of a kind between different species of plants. Association can be a loose form of what biologists call "symbiosis," though it encompasses few species rather than many.

But this kind of association is a wholly Darwinian idea. An association between pairs of species comes about as each is separately programmed to achieve fitness. Practical experience now shows that dependent relationships of this kind are usually loose and never extend to many species at a time. We can assert this because there has come a time in the studies of communities when every patch of vegetation that was claimed as an entity or plant society has dissolved before critical inquiry. This was true first of the continental formations, the nation states of vegetables, which had boundaries only where there was a physical boundary on the earth—a coastline, a mountain range, or an air mass. Then it was true of the "associations" and "sociations" of the phytosociologists, all of which could be shown to reflect drainage, or exposure, or soil. And it was true of life-zones on mountains, which turned out to be simplifying abstractions like a band of color thrown by a prism of glass, which always blends imperceptibly at the edges.

There are no discrete communities of plants. The reality is endless blending as each individual Darwinian species finds its own range, jostling its neighbors, living in its own individual niche. Communities come and go, mere temporary alliances of plants thrown together by fate and history. But while they live together their livelihoods mesh with those of their neighbors. The realization of this fact has been the biggest fruit of plant sociology. The botanists who looked for organized plant societies in nature gave us an entirely new concept of how life actually was lived and called this concept the "ecosystem."

In the 1930s botanist after botanist made his own private pact with reality, recognizing that his "community" was really defined by a patch of ground and that the fates of its members depended on soil, weather, and the animals of the place more than they did on the lives of neighborly plants. Then the word inventors went to work. There was the *naturcomplex*, the *holocene*, and the *räume*. We read of *biotic districts*, of *biochores* and of *biogeoceonoses*. An idea had found its time and wanted a word that was clear, expressive, and did not twist the tongue. It was found when the English botanist, Tansley gave us the "ecosystem."

The ecosystem was thus an invention of botanists. I use the word *invention* with care. The ecosystem describes an idea, a people-made thing. The idea was that patches of the earth, of any convenient size, could be defined and studied to see how life worked there. The physical and the living must be looked at together to see how each acted on the other. The ecosystem concept is a way of looking at nature. It is an admission that there is no super-organismic thing out there made by some masterly designer. There are only Darwinian species. But if we would know how these persist on earth we must look at both niche and habitat in a systematic way. We must study both the quick and the dead.

Tansley was the kind of scholar that other scholars revere, learned, wise, and moderate. He left behind a masterly treatise on the vegetation of Britain, and his king knighted him for his services to English botany. Possibly King George did not know that he was honoring not just a great scholar but the inventor of the ecosystem. The knighthood of a plant-hunter might have brought a chuckle to the lips of they who wore the golden spurs at Agincourt or Crécy, but it is not hard to see who left the greater mark on his posterity

# Chapter Seven. Cycles: A Lesson From Farming

MANY European farmers have found out the hard way that the methods of their ancestors do not work in parts of the wet tropics. If you clear a typical piece of red tropical land, plough it up, and sow seed, your efforts are poorly rewarded. There may be a few years of struggle against falling yields, but then comes the bitterness of defeat, and a patch of red mud is left for the wild weeds. This pattern was so well-known to indigenous peoples of tropical places that they used to expect to abandon such land after a few years, and to move on to clear another little patch, a proceeding known as "shifting agriculture." Continuous-yield agriculture of the kind that has sustained Western civilization does not seem to be possible in many tropical places, and this is, on the face of it, a very strange thing. The tropics are warm, and surely crops like warmth? Many of the places where farming fails are wet, and yet plants like water. And, what is more, the wild vegetation of these places may be luxurious almost beyond a northern farmer's imagining. Yet when he tries to farm this verdant place, he fails.

The immediate cause of these failures is well-known. These tropical soils are infertile in the chemical sense; they lack phosphorus, potassium, nitrogen, and lime. The soil is warm, and has been soused with splashing water since the dawn of tropical time, a proceeding that might very well wash potassium, calcium, and the rest

out of the ground. Indeed, the endless sousing with
warm water is part of the reason the tropical soils are
red, for the rains have taken out not only the soluble
plant nutrients but even silica, the flinty heart-substance
of rock that gives the gray-tints to northern lands. All
that is left in an ancient soil of the wet-tropics is an insoluble matrix of the simpler clays with aluminum and iron
oxides, a red mixture well on the way to becoming baux-
ite or iron ore.

So the Western farmer fails in many tropical places
because the soils he ploughs have been washed clean of
plant nutrients. But how does the wild vegetation get
along without these essential chemicals in the ground?
Plant nutrients are as important to a rain forest tree as
they are for wheat or corn, and the land may well have
supported a massive forest of these trees before the
farmer cleared it. How have the forest trees, with their
associated hosts of other plants that we collectively call
"the tropical rain forest," managed to thrive in a soil
washed clean of nutrients. The answer is to be found by
looking at one of those forests before a well-meaning
farmer cuts it down.

Where the forest is intact and healthy, there is a
goodly reservoir of nutrients, in spite of the state of the
soil. The supply is in the living plants themselves. In this
ecosystem, the living things are their own reservoir of
nutrients, and the forest is responsible for maintaining
and replenishing its own chemical reservoir. It is on the
face of it an uncertain reservoir because it is always being
raided. Whenever a great tree dies, its corpse is torn
down by fungi and termites, and its special portion of the
communal reservoir is thrown down with its rotting parts
onto that red washing-board of the mineral soil. All the
time animals are biting away at the green leaves and
dribbling a leak of nutrients down to the soil in their

excrement. But these lost nutrients are not allowed to get clean away, for the forest retrieves them all with an extremely complex and elaborate network of the finest roots, and with the help of special kinds of fungus that live on those roots. Almost nothing of value to plants gets past this network to be washed away in the drainage water. This is the secret of how the rain forests thrive on infertile soil; the living community hoards and recycles the chemical nutrients needed for life.

When a farmer clears a lowland rain forest, he kills the plants, destroys the retrieval system, and lets the chemical nutrients be washed through the soil, into the rivers and down to the sea. When his demanding crops fail, he says the soil is infertile. Of course it is, he made it so.

When the farmer, impoverished by his folly, quits, wild plants will grow on the red ugliness he made, but not the forest he removed. Specialized plants arrive that can make do on extremely short commons for nutrients. With luck, these will begin the process of hoarding what little is left, more nutrients will slowly collect from rain and dust, the reservoirs will slowly be built up again, and one day the forest with its ample magazine of supplies will be restored. Or so we may imagine, because we can only guess that this might be so, speculating that something like it must surely have happened to put the forest there in the first place. But hundreds or thousands of years might be involved. Sometimes a tropical rain forest may have originated in ancient times when the climate was different or the soil young, and merely persists in our own day by reason of its living cycles. When we destroy such a forest as that, it may be gone forever.

We now need to know why the conventional agriculture of Western man works anywhere, even in his native Europe. The best farmlands of the north are well-watered, being regularly soused by rains quite capable of dissolving potassium salts, nitrates, and the rest, of washing them through the soil and carrying them to the sea. Yet when the wild forests of these lands are killed, and the brown soil ploughed, crops can be grown year after year. On some of these soils we have been at it for several thousand years and we have not been forced to quit yet. It is true we have learned that it is a good idea to put the excrement of our animals back on the land, or to replace this with chemical substitutes, but even bad husbandmen get something. We never suffer the retribution that awaits he who takes these methods to the red tropic earths. This, for all that it is common experience, is really rather odd. Why should these drastic proceedings of agriculture work at all?

Understanding the nutrient reservoir of a living temperate forest gives us the answer to this. In the temperate forest we find that there are three reservoirs of nutrients instead of just the one found in the tropical forest. The first reservoir is that same magazine in the living plants themselves, but this is only a small part of the total. The other two parts are held in the soil humus and in the curiously complex clay minerals of the colder soils.

The usefulness of soil humus is obvious, and so is the reason for its existence. In the colder northern regions, with their freezing winters, the process of rotting is much hindered, so the more resistant parts of dead plants last a long time. They yield to the soil water the nutrients that went into their making only grudgingly and this trickle can be captured by root-systems that are a good deal less efficient than those of tropical trees. Even crops can intercept this trickle. Furthermore the chemistry of humus is such that nutrients such as nitrates and phosphates, those that are negatively charged, actually collect on the humus particles, being dragged out of the soil-water by inanimate chemical process. So one of the reasons that temperate agriculture works is that the cool climate lets humus collect, and the humus acts as a passive, non-living regulator of the nutrient supply.

The different collection of clay minerals may be even more important. We see the physical difference when we see that the tropic earth is red, unlike the northern earth, which is brown or gray. For plants it means that complicated aluminosilicate minerals, with tongue-twisting names like montmorillonite, are not present in the wetter and ancient tropic soils. These aluminosilicates are the sticky clays of the north, the soils that suck one's boots off crossing a wet, ploughed field. But they suck and hold nutrients as well as boots, for they are negatively charged and collect metallic ions such as potassium, sodium, and calcium at their surfaces, providing a store on which the plants can draw. It is this storehouse of nutrients on aluminosilicate clay minerals that is the third reservoir of fertilizers provided for temperate soils.

If we are to understand fully why Western agriculture works much better in the north than in typical tropic lands, we must first understand what makes a soil either red or gray. It has not been easy to arrive at a firm answer. There is a correlation with temperature, but a correlation does not imply direct cause. Red soils are made where it is warmer, but why should warmth make for a redder soil? Other correlations are possible, with the kinds of plants of warmer places for instance, or with soil animals, rather than with straight temperature. The most interesting correlation was once offered to me by a native Virginian as we drove south from New England to the Carolinas. As we passed Virginia we went through the transition from the grays and browns of Yankee land to the reds of the South, a transition as exciting to human emotions as it is intriguing to the scientific mind. It moved me as it usually does, and I remarked on the redness of Virginia to my companion. "Aw," he said, "the ground is just stained with Yankee blood." Virginia soils are only a little red. They still have silicates and are good for farming, unlike the redder soils of real tropical lowlands. But perhaps Yankee blood is the fertilizer.

The silica of the aluminosilicates is missing in the red earths. In northern soils the perpetual washing of the rains leaves many of the flinty-gray silicates of the old parent rock behind in the soil. In the coldest of well-washed soils, the podzols of the northern coniferous forest, a bleached layer of silicates remains as a residue near the surface. In the browner soils of the deciduous woodlands, the soils in which we do most of our farming, there is still a rich residue of silicates, though mixed and combined with oxides of iron and aluminum. In the tropics, however, the silicates have been washed away. Without silicates, the paths of mineral synthesis in soils go different ways, and tropic soils are deprived of the complex clays that hold on to nutrients. Both the problems of the red and the gray and the poor fertility of tropic soils then relate to the mechanism that moves silicates when it is cold but leaves them when it is a little warmer.

The chemistry of a soil is so complex that accurate descriptions of this mechanism have not yet been made. It is well to think of a soil as an immense filter bed, each tiny particle of which is chemically active. Water trickles through this maze of reaction sites, water whose acidity is dependent on the minerals it has already passed, on the organic rubbish through which it has percolated, and on the gases that are breathed by the life of the soil. Acidity affects the rates at which different minerals are dissolved, but so does the chemical milieu of other minerals or the presence of organic colloids. The synthesis of new minerals, always proceeding in a soil, in turn reflects on the concentrations of solutes in the soil solution, which in turn controls acidity and fresh synthesis. There is an immensely complex chemistry going on in all soils. Part of it, perhaps particularly the acidity of soil water, is powerfully affected by plants and animals. Other parts are the results of purely physical processes. And all processes, whether living or dead, are affected by temperature. Whatever the detailed mechanisms may turn out to be, it is clear that in warmer climates the chemistry goes to the exclusion of silicates whereas in cooler places the silicates are saved and synthesized into minerals such as montmorillonite, which make the soil gray and do much to make it suitable for agriculture.

Western agriculture is successful in its homelands because northern soils hang on to nutrients with only little help from the plants. This is a physical accident, an indirect consequence of the coolness of the weather. A large part of the nutrient cycles of a northern ecosystem is worked by non-living processes, of which the plants and animals take simple advantage. Farming man cannot kill what is not alive in the first place, and so his replacing one vegetation with another does not destroy the essential life-support system of the place. He can, of course, be so crassly careless that he lets the soil erode away, but he usually learns not to do this. With the soil intact, the magazine of nutrients is largely intact also.

There are some places even in the tropics where there is a large non-living input to the nutrient supply, the most obvious of which are the great river deltas. At the mouths of the Ganges, the Niger, the Mekong, and the Nile masses of sediment are dumped every year, and with this sediment comes many of the nutrients that have been lost by all the ecosystems upstream. You can work this land as hard as you like without spoiling it, because the physical floods will mend the damage every year. Which is why the estuaries of tropical rivers so teem with people (until the rivers are dammed).

There are also other places of fertile soil in the tropics, particularly where soil has been made out of young, nutrient-rich volcanic rocks. It is these soils that have made millionaires out of pineapple growers in Hawaii. Other fertile land is, strangely enough, in places where the rainfall is reduced, because then the soil chemistry is different. The drier half of the Indian subcontinent has black soils manufactured by heat, wet, and a dry season, and these are good soils for farming. But in the typical lowland tropical place where it is always hot, and where the rains have swept down on an ancient land surface for thousands of years, the soil has almost no nutrients, and vegetation only persists by building its own magazines. The plants of the wild ecosystem have been made by natural selection to do this. When you kill them, you destroy as well the system that supports life.

# Chapter Eight.  Why the Sea Is Blue

THE sea is blue. This is a very odd thing because the sea is also wet and spread out under the sun. It ought to be green with plants as is the land, but it is not. There are murky coasts and estuaries, the green hard waters of stormy channels, the fog-covered silvery-gray of ocean banks. But the deep sea, the open sea, most of the sea, is blue. This strange blueness of the sea can tell us many things.

An explanation for the color of the sea is simple enough. There are not enough plants in the sea to make it green, so we are left with the color of pure water under the sun. Light that passes through perfectly clear water is absorbed bit by bit, its energy dissipated as heat as it travels until at last all of it has flowed into the sink of heat, and there is an utter blackness. But the colors of white light go progressively, one at a time. The low-energy wavelengths that we call "red" go first, then in turn the more intense parts of the spectrum—orange, yellow, green—and finally the various shades of blue. Only blue light reaches a few hundred feet down, and it follows that any reflected light that has made a double journey from the surface to the depths and back is blue. And so the sea is blue.

But the real reason that the sea is blue is that there are not enough plants in it to make it green. And this is one of the oddest of the odd things about our world. Why are the great oceans not green with plants?

We can get a first hint of where to look for our answer by reflecting on those few places in the sea that in fact are green, in particular the shallow banks such as the Dogger Bank or regions of great upwellings such as those on the Peruvian coast. These are the sites of the great fisheries, and the waters are murky green with plant life. The fisheries themselves attest to the rich productive qualities of these scattered places, and the green murk bespeaks high fertility. Indeed, high fertility, in the simple chemical sense, is the explanation of both the fisheries and the green murk. The waters of banks and upwellings are well-supplied with chemical nutrients, so that the tiny planktonic plants of the sea thrive abundantly, turning the water into a green soup in which animals wallow, to the eventual well-being of fishermen.

Where the sea is unusually fertile, tiny plants multiply and the water becomes green with their bodies. But most of the sea is not fertile; it is a chemical desert. Potassium, phosphorus, silicon, iron, nitrates and the rest are always present in sea-water, but in low concentrations. By the standards of agriculture, the open sea is hopelessly infertile. And if the sea is infertile it is perhaps not unreasonable to expect that plants will not grow there very well, which is presumably why there are so few of them.

So far we seem on safe ground, but there is a very large catch to the argument. Everything depends on the fact that the plants of the sea are tiny. In very fertile water (a polluted estuary is the best example) the tiny plants multiply until the water is pea-green with their bodies. But if the water is a nutrient-poor desert like the great oceans, then there are only enough chemicals in the lighted upper layers of the sea to make a very few plant cells. The water is then essentially empty, the sun plunges down to the depths, and the water glows blue. But all this only follows *if* the plants are tiny.

Suppose there were large plants floating on the surface of the sea, plants that covered it with layers of leaves as the rain-forest trees cover the tropical land. These large plants would not have to worry about the thin supply of nutrients in the water any more than the rain-forest trees of the last chapter are stopped by the even thinner supply in the red tropical soils. Large plants can collect, accumulate, and hoard nutrients. How easy it ought to be for a large plant in the sea. Deep down below the lighted surface of the sea there are, in fact, almost unlimited nutrients, for the great oceans are some five miles deep in the middle. The fertilizer problem is one of concentration. In the few tens of meters at the top of the ocean, where the light reaches and the plants must grow, there is a local shortage of nutrients, but the potential supply underneath is truly enormous. A large plant at the surface would soak up nutrients, just like a large plant on land. More nutrients would then diffuse up from the depths to be similarly collected. And so on. Thus, if there were large plants in the open sea, the dilution of nutrients would not matter.

Now our inquiry comes close to Darwinian realities. The sea is blue, not so much because it is actually infertile, but because there are no large plants growing there. Large plants would overcome the infertility of the surface water by gradually collecting nutrients as they filtered up from the depths below. Large plants would become the dominant factor in the life of the sea as they are of the life of the land, making massive shade of the spaces below them, forcing all food chains of animals to start with types that could bite out chunks from foliage. But no large plants live in the open sea. They can grow around coasts as do the kelps. The giant kelps of the American Pacific, *Macrocystis* and *Nereocystis*, are said to be the longest vegetables in the world. But none of these large sea plants makes it out to a floating life in the open sea. For some reason the niche or profession of "large-planting" is not possible in the open sea. Why? This is the fundamental Darwinian question behind the blueness of the sea.

Oceanographers have long known that there is something odd about the absence of big plants from the sea, but they have missed the grand Darwinian question. They never asked themselves, "Why can't the plants be big?" Instead they looked for the advantages in being small, counting the blessings of smallness and expecting to find their answers in this way. But you cannot get all of the answer like that.

Consider some of the so-called advantages of being small, particularly those based on surface area. A small object has a much larger surface in proportion to its volume or mass than a big one. One result of this is less trouble with the sinking-problem, since the relatively large surface offers more friction, slowing the sinking rate. On the other hand, if you have a bladder of air or oil, you do not sink at all, so why bother being small? Another is that the large surface to your little body can be used to soak up scarce nutrients. But there are ways of having a large surface area other than being tiny; by being convoluted or sponge-like, for instance. Rain-forest trees manage with a mat of hairs, even in mud and gravel let alone water. A spongy giant of an ocean plant would find soaking up nutrients easy; then it would be able to hoard its nutrients even as land vegetation does.

I have read in an oceanographic text that small entities use nutrients "efficiently." This means that "turnover" is efficient, if one thinks of the oceans as a banker thinks of a company that "turns over" its capital quickly. But it is a strange sort of efficiency that keeps the oceans as a poorly productive desert. If the ocean plants were large, they would soak up nutrients from below and make the ocean desert bloom like the lowland tropics. The "efficiency" of production would then be much greater. So why be small?

There must be some advantage in being small, and we can best find what it is by looking for the disadvantage in being big in the sea, and whatever this disadvantage might be it surely must be overriding. There are big plants everywhere else, on all kinds of land surfaces and in every shallow patch of the sea along its coasts. It is only in the open sea, where they would have to float that there are no big plants. So the answer to the problem must lie in the floating way of life.

Why do small plants make a success of floating in the sea whereas big ones do not? The answer stares us in the face. If a plant floats, it drifts, and if it drifts, it is soon blown away from where it wants to be. There must be some way to get back. A big floating mass, kept up by air bladders or oil floats would never make it home after the first storm or the constant push of a current had taken it away. But it is easy to imagine ways in which tiny plants might arrange for their returns. The most obvious way is by letting themselves sink, because the surface of the ocean is always being stirred. Water always moves into a patch of sea as fast as water is taken away and for every leaving current there must be a current returning. It seems likely that small plants can thrive in the open sea by following the currents round. It is possible, too, that they disperse in the air as well, being kicked out of the waves in spray and blown about the world oceans. Tiny plants can ride the currents to stay in home waters or travel the oceans to get back there. Large floating masses of vegetation cannot.

So my final hypothesis to explain the blueness of the sea is that large plants are excluded from it not by short

commons in nutrients, but by the restless motion of the
waters that would sweep them all away never to return.
As it happens, fate has provided one intriguing test for
the hypothesis in that there is one place in our contemporary ocean from which floating things are not swept
away: the Sargasso Sea.

The Sargasso is at the center of a slow but enormous
gyre, an oceanic whirlpool that gathers floating debris to
its middle. This was so dangerous an area for sailing
ships that legends have grown of ancient vessels, trapped by the remorseless swirling waters, rotting together
far out in the Atlantic. Columbus had his own grim meeting with the Sargasso, saving himself from the mutinous
temper of his crew only by scooping a crab off the weed
that floated alongside and claiming that the weed with its
crab meant that land was near. But the land was a long
way yet. The weed was the big floating brown alga we
call *Sargassum*, and it floated thickly over the surface of
the Sargasso Sea because the gyre held its population in
place.

Sargasso weed floating about as straggling fragments
can be found in many of the world's oceans as can fragments of other species, of *Fucus* and *Ascophyllum*, of
any of the anchored coastal plants that bear floats and
that might be torn up by storms. These floating fragments survive for a while as they drift, but they are all
doomed. They are not adapted to oceanic life; they cannot reproduce as they float about; they leave no offspring; and they die. But in the Sargasso Sea things are
different. There the local species of *Sargassum* lives its
whole life, reproducing, and persisting generation after
generation. Evidently this gyre in the oceans has per-
sisted long enough for natural selection to produce from
the chance debris of floating coastal algae a species able
to carry on its life cycle as it floats. And the plant has

done this in a patch of water notoriously unproductive in the sense of holding few nutrients.

The story of the sargasso weed leads us to believe that where it is possible for floating plants to stay put in the sea, we shall find large, floating plants. That we do not find them all over the oceans is because the oceans do not keep still. Natural selection then forces extreme smallness on the plants that are there, for the tiny ones are those best able to disperse about the seas. If the surface waters are provided by conveyor-currents of nutrients in upwellings, or run-off from the land, or with delicious rivers of garbage like those that pour from the Tiber, the Hudson, or the Medway, then the tiny plants will so multiply that the blue of the ocean is banished, and a green or turbid murk tells of vibrant life.

But if the sea is a nutrient-poor desert, like most of the world oceans, then the tiny plants cannot be very numerous. There is then neither a canopy of floating vegetation nor a soup of tiny algae. Sunlight plunges deep into the water, its less intensive rays being rapidly extinguished the while. Only the shorter wavelengths make the double journey to and from the depths. Which is why the sea is blue.

# Chapter Nine. The Ocean System

THE world oceans make up a vast desert, desperately short of nutrients and with living things spread most thinly through them. This is the shocking message of our inquiry into the blueness of the sea. I use the word "shocking" with care. Our generation has been treated to tales about the sea as the last frontier, as a place of wealth, of riches, of production. No journalist seems to be able to write on the subject of feeding the hungry without mentioning farming the oceans, as if they were some great untapped source of food for people. But they are not. The oceans are deserts with little more food in them than we are taking out already.

I described in Chapter Four how we can measure the efficiencies of plants by weighing and by totaling all the new tissues they made in a growing season or by monitoring the gases they took in or breathed out. We could measure in that way just how many food calories were made by the plants of a piece of land during the course of the year, and we found out how lamentably inefficient land plants were. Things are far worse in the ocean.

It is easier to measure the productivity of the seas than of the land because the plants live in water (an ideal medium for the chemist) and because their small size permits a nice population of them to be contained in a laboratory-sized bottle. The oceanographic ships of many nations, which now patrol the world oceans, routinely drag up bottles of sea water to look at what the plants are doing, and we now have good measurement from all the seas of the world. The results are grim.

All the seas of the world taken together produce about 92,000,000,000 tons of plant tissue a year, a figure that includes the fertile places with celebrated fisheries as well as the blue waters of the tropical oceans. This may seem a large figure but it needs to be compared with what the plants of the much smaller area of dry land can do. The gross production of all the plants of all the dry land of the world is about 272,000,000,000 tons of plant tissue a year. And so we see that, although sea water covers nearly three-quarters of the surface of our planet, the plants in the sea account for only one-quarter of the calories fixed by living things.

The immediate reason for this appalling unproductiveness of the sea is, of course, the scarcity of chemical fertilizer. On land, fertilizing nutrients can sometimes be in short supply too, though wild vegetation usually manages to hoard and cycle the stuff to meet its needs. As a result, unless there is no water, or it is winter and too cold, the production of land plants is set by shortage of the raw material carbon dioxide. In the sea, plants get their carbon in solution as the bicarbonate, and usually they get more than they can use of it. And the reason the tiny sea plants cannot work up to *their* carbon limit is that they run out of chemical fertilizers such as iron, phosphate, or nitrate first.

A first reaction of technically minded man to the infertile sea is to fertilize the wet desert and make it bloom like a soggy rose. But this will not do. It is not that there is an actual shortage of chemical nutrient in the world oceans, because the actual reservoir of chemicals it contains is enormous. Indeed, we sometimes dream of "mining" the oceans for the minerals its waters hold. The problem is one of dilution. Plants must live in the top layer of the oceans where sunlight enters down to say a hundred meters (about three hundred feet) but often much less than this, and the only nutrients of much interest to them are those in this top, thin, lighted layer. All the tonnage in the vast deep beneath is out of reach of the plants. If we dumped more millions of tons of superphosphate and ammonium salts into the sea (assuming we had them), they would merely fall to those same inaccessible depths.

But we know that nature herself has made some patches of the sea fertile, for this is where we go fishing —the North Sea, the Newfoundland Banks, the waters off Peru—and it seems strange that some patches in this fluid thing, the sea, can be fertile while the rest remains a desert. But the explanation is simple enough. The waters that support the fisheries are not really all that different chemically from the desert waters, but they are continuously replaced. Where there are shallow banks or island arcs, the moving currents of the deep sea are forced to the surface, so that the water flows up from below. A similar thing can happen when two deep currents meet head-on so that water is forced upward, leading to some of the celebrated "upwellings." In all these places of banks and upwellings, life is provided with a vertical, slow-moving conveyor belt of water that brings endless supplies of fresh nutrients from the depths. Continuous replenishment of the thin nutrient broth at the bubbling head of a current is the secret of a productive fishery.

There are also fertile inshore waters—the narrow coastal strip or the turbid sea off the mouth of a great river. To these places good fortune comes in two ways, both because the slope of the land forces currents to rise, so that they convey nutrients endlessly as they break against the shore, and because the edge of the sea is fertilized by the debris the rivers bring. How important the fertile flow of a river is can be shown by stopping its flow, an experiment recently tried with the Nile. The Aswan dam has prevented the Nile from discharging its nutrient-laden silt into the sea, and the immediate consequence of this has been the collapse of the sardine fishery as the plants, which fed the plankton that fed the sardines, failed to grow. We have also sometimes been able to increase the amount of fertilizer in the sea near coasts and rivers to levels to which the local plants are not accustomed. The result is often called "pollution."

But, apart from narrow coastal strips, the only fertile patches of the oceans occur where currents from the cold depths bubble to the surface. Plants do not like the coldness of this water, but they put up with it for the sake of the nutrients it brings. There the little plants can live their short lives, bobbing in eddies so that a parent stock remains near the middle of the upwelling, passing through their fifty or sixty generations a month, stoking the food chains that support our fisheries. Only one tenth of one percent of the world oceans is a place of true upwelling, and about 10 percent is moderately productive coast. The immense remainder is a blue desert more useless to life than most of Arabia.

Thus, the fact is established: the oceans are cruel deserts, and the things they lack are the soluble plant nutrients. But this is a strange conclusion for one who broods about the history of the earth. The oceans have existed since almost the beginnings of earthly time, changing in shape, shoved from one part of the planet to another before the drifting continents, but always present in roughly the volume we know. And all the time soluble nutrients have been washed into them by the rivers coming from the land. The sea has been made salty by this process. And yet it lacks the nutrients needed for life. Strange.

Some details of the answer to this riddle still escape us, but the outlines are clear. The chemistry of the sea is controlled by its mud. Even as the endless-flowing rivers discharge their chemicals into the ocean, so the mud at the bottom soaks them up. The mud is selective. It has complicated minerals similar to the clays of temperate soils that hold metallic cations like calcium, potassium, and sodium. In its medium of salt, the surface of the mud allows slow crystals to grow, like the nodules of manganese that some mining corporations plan to dredge for. There are sites where calcium carbonate precipitates and collects into reefs of limestone, dragging with it other elements such as magnesium. The mud contains organic debris on which bacteria do their strange feeding, fixing some elements to their corpses and discarding others. In these ways the chemistry of sea water is warped away from the typical chemistry of the water of rivers, principally by chemical reactions on its bottom. The sea is not only more concentrated than fresh river water; it contains a quite different mixture of chemicals. This strange mixture is largely fashioned by sedimenting minerals and its mud.

Chemicals are removed from the ocean basins as fast as they arrive from the rivers. They flow back to the land with the writhing of the earth's crust. Every thrust of a mountain range and every emergence of a coastline from the sea brings the chemical-rich mud back to the land. All the sedimentary rocks, from limestone and sandstone to shale and schist, were once part of the ocean's mud. When they were lifted out of the sea, they took with them the chemical nutrients stored in them. At once these nutrients began to be washed out of the rocks again by rain, to be caught in the roots of land plants and held for a while in land ecosystems, then to escape in a slow leak to the rivers, and to be sent on another journey to the sea, another spreading through the fluid mass, and another sorting on the sea's bottom.

It is an enormous chemical machine that keeps the sea a desert. All the chemicals in the sea are cycled slowly through it. They come from the rivers, they spend a time in the oceans, diluted and in suspense, then they are taken in by the mud, held for a brief few million years, and then thrust back onto the land in a prison of rock. This is a system that holds the chemistry of the sea constant from eon to eon. It is not an ecosystem, for all that bacteria and other forms of life do some of the chemical things on the way, particularly influencing the deposition of carbonates. It is a passive physical and chemical system, driven by the sun because the sun is there, but not organized by life.

To live in the open sea, one must adjust to the lack of nutrients. A plant cannot be large for the reasons discussed in the last chapter. The niches of small-planting in an unproductive ocean require that plants spend much of their time drifting in the lighted skylight of the sea. They can sink a bit into the dark, shutting down their factories for a while until some eddy brings them back up, and this helps them and their descendants get about. Their breeding strategy is forced on them by their size; they must swell and divide as fast as possible. And they are horribly vulnerable to the grazing animals that hunt them through that glowing open place.

An animal who would adopt the profession of hunting down these tiny plants faces a similar set of constraints. Either it takes the plants one by one, hunting them as a trout hunts a mayfly or a fox hunts a mouse, or it must filter the tiny things out of the water with some sort of sieve. The way the herbivore chooses its size is set by simple principles of mechanics. It must be very small and, as a direct consequence of this, constructed on tolerably simple lines. There can be no brain for cunning, no elaborate eyes to show proper pictures. It hunts through the skylit void according to simple mechanistic rules. Thus we have the herbivorous copepods and their kind, perhaps including animals up to the size of the krill that whales eat.

These "choices" of which I talk are, of course, made by natural selection. There is no conscious design or free will, but the options have been set by the size and habits of the plants and we can see what they must be. Once these options have been taken up, we can see what new possibilities open for natural selection. The first is the flesh-eating animal that will hunt down these humble hunters of the plants. He too must pick his way through the void, seeking his more scattered prey, needing better hunting techniques, being larger and more complex. And he must face the hard reality that natural selection will have provided fiercer animals still to hunt him through the lighted open places where there is no cover and nowhere to hide.

Fishes high on food chains are colored silver beneath and dark on top, no matter from which sea you take them or from what ancient stock they have evolved. Very clearly they are hunted (or hunt) animals that use eyes to locate their prey. Many of the larger animals spend their days deep in the sea where there is no food but where enemies cannot find them either, only coming to the surface at night to hunt the plankton and smaller fish herded there where the plants live.

Thus ecosystems build in the sea into patterns quite different from those of the land. These oceanic ecosystems can do much less to modify the harsh facts of physical existence. Not for them the regulation of the nutrient supply, of providing shade or physical structure and hiding places, as is done by the living things of a terrestrial forest. For nutrients and dwelling places, the life of the open sea must make do with what is provided by passive physical and chemical systems. What we have instead of regulation of the physical habitat is an ecosystem of hunt and be hunted, where plants and animals are ruthlessly adapted to living in a nearly desert but brightly lighted lens of water floating over a black immensity.

When people enter the sea in quest of food, they harvest the produce of a desert, which means that they cannot get much, however large the desert may be. But matters are even worse than this implies because people cannot get at those tiny plants. Nor can they get at the small animals that eat those plants. For the most part, they cannot even get at the animals that hunt the animals that eat those plants, and must fish further up the food chains. With each link of the food chain, as we have seen, some 90 percent of the food calories originally present are burned away.

When we fish in the oceans we do not just harvest the meager produce of a desert. We get instead (when we are clever and lucky) some 10 percent of 10 percent of 10 percent of the harvest of a desert. The best estimates suggest that we are already fishing close to what these sums say the oceans can bear.

# Chapter Ten. The Regulation of the Air

THE prelude remarked on the uniqueness of the gassy mixture in our atmosphere. Nothing like it exists anywhere else in the solar system, the other planets of which have suffocating heavy mists of hydrocarbons such as methane, or of ammonia gas and carbon dioxide, or they have desperately thin miasmas of these same poisonous brews. But we have oxygen flowing round us with which to work the chemistry of our lives, and the oxygen is buttressed by a grand dilutant of inert nitrogen. Within these two are balanced the vital low concentrations of the combinants of oxygen: carbon dioxide and water vapor. It would have been very hard to predict the existence of such an odd mixture of gases at the surface of a planet floating in space had we not already known that they were there. And we have good reason to suspect that life itself has something to do with the existence of this strange mixture, and with keeping it in good order.

Consider the nitrogen first. If we lost a significant portion of that inert dilutant, everything around us would probably go up in flames. What new chemical reactions would then be set in train among the remaining gases of the air I am not chemist enough to guess at, but there would not be much in the way of macroscopic life remaining to study the process. We need inert nitrogen; inert though it is, there are living things that are busy taking away nitrogen from our air. They are called "nitrogen-fixing bacteria."

We first learned of the existence of nitrogen-fixing bacteria from the experience of farmers. Long ago it was the custom to grow crops on the fields for two or three years, and then to leave the land alone for a year, a practice called "fallowing." Farmers who went on cropping year after year without a break soon found their yields were low. The land produced a good crop only for two or three years, then it seemed tired of being extorted, and it struck. But all the encouragement it needed to start producing again was a year's rest. If the land was left fallow every three or four years, good crops might be expected from it indefinitely. The farmer reasoned that he had to "give it a rest sometime," and this simple analogy to human labors let him practice good land management. One of the most important things that happened during that year of rest was that nitrogen-fixing bacteria took nitrogen away from the air, combined it firmly with oxygen to make nitrate, and bequeathed this nitrate fertilizer to the soil.

Nitrogen-fixing bacteria are some of the great servants of mankind, and in all the world's high schools children are taught about the root-nodules of legumes, the peas, beans, and soya, where these friends of ours live. Nitrogen-fixing bacteria make nitrogenous fertilizer so that crops may grow. They are, however, constantly removing nitrogen from our air in such a way that, were their labors not undone, they would end life on earth.

Nitrate is a stable substance. There is no obvious reason why it should not collect on the surface of the earth in great mountains as does its chemical relative carbonate, which, in its limestone form of calcium carbonate, gives us the White Cliffs of Dover and many a mountain range elsewhere. True, nitrate is rather more soluble in water, but this should merely result in nitrate reefs in the sea similar to the carbonate reefs there now. The logical result of the efforts of those friendly nitrogen-fixing bacteria should be that both the nitrogen and the oxygen would be taken from our air and piled into white mountains on the land together with white crystalline reefs in the sea. The miasma of gases left overhead would be like the thin evil brew on Mars. As far as we can tell, we have been saved from this fate only by the activities of other bacteria.

These other bacteria live in stinking mud; in marshes, at the bottoms of polluted lakes, and in the sediments of fertile estuaries. They feed on the rich supplies of fuel that collect in these places, on water-logged organic matter, and on various chemical fuels associated with the organic refuse. Indeed, the black mud of bogs and lakes is potentially a very rich source of stored food for an animal whose chemistry is odd enough to let it live there, but there is a great difficulty to be overcome. The boggy bottom is without oxygen, and without oxygen an animal cannot burn these organic fuels to make heat and do its work. The solution found by natural selection for some of the bacteria of these places is to take the oxygen from nitrates.

There are many ways in which this trick of breaking up nitrates is done in those boggy bottoms, many of them involving the manipulation of sulphur compounds which gives the characteristic smell of hydrogen sulphide to marsh mud. But all the mechanisms work the same essential trick: they borrow oxygen from nitrates so that nitrogen eventually is freed from its oxide bondage and allowed to bubble back to the air. The oxygen too gets out in the end, most likely as carbon dioxide, but perhaps as carbon monoxide or some less usual gas. And the result of the activities of the bacteria of evil-smelling mud is that they replenish our air. Without their work, the life-system of the earth would run down and stop.

The bacteria who give us back our nitrogen are also important producers of oxygen, since the workings of their lives let the oxygen of nitrates leak back into the atmosphere. But green plants are far more important producers of oxygen. They make oxygen all the time they run their solar-powered factories because they make sugar out of carbon dioxide and water, a process that always results in some oxygen. So all the green life of the earth is steadily pumping out oxygen, and all the plants together pump out vastly more oxygen each year than do the bacteria of swamps.

Green plants may well be the most important producers of oxygen, but we have detected one physical process that is helping. High above the atmosphere, where the more deadly of the sun's rays still play, water molecules are smashed into oxygen and hydrogen. The light hydrogen gas is lost to outer space, and the heavier oxygen remains, but we do not know how much oxygen we get each year in this way. The gut reactions of many a scientist who has thought about it is that plants are much more important. Indeed, our best models suggest that the early earth had no free oxygen in its air and that the first plants put the oxygen there. The very existence of the remarkable air of our planet thus becomes one of the more spectacular of the activities of life itself.

So we come to the realization that the remarkable air of our planet is regulated by the life around us. Maintaining the air is one of the joint functions of all the ecosystems of the earth. And then we reflect that we are now tampering with those ecosystems on a grand scale. What if we accidentally impede vital parts of the system of maintenance? Loss of our oxygen supplies has been the most spectacular of the retributions promised us by those who have warned of environmental catastrophe unless we mend our technological ways.

In its simplest form the scenario of oxygen catastrophe relies on poisoning the sea. The argument says that we have shipsful of herbicide sailing the world oceans, that one or more of these might sink, and that the herbicides might spread through the sea and kill many of the sea plants. All this has a nasty plausibility and makes any person who loves our earth want to make sure that ships loaded with herbicides be prevented from sailing the world oceans. But the argument goes on from there to say that the loss of these sea plants would result in that critical impedence of the mechanism for regulating our air. With this threat the argument errs.

It used to be believed that much of the world's plant production was in the sea, and some elementary biology textbooks still talk about 70 percent of all photosynthesis being in the sea, or even 90 percent. If this were true, then killing all the plants of the sea might stop more than 70 percent of our vital oxygen pump, which sounds a rather dangerous thing to do. But most of the plant production of the earth does NOT take place in the sea. The seas are those cruel deserts of which I wrote in the last two chapters, and together they account for 25 percent of plant production of the earth, not 70 percent. It is not likely that even the worst excesses of our folly could kill the plants of all the seas. But, if we did, we could knock out a quarter of the annual oxygen return while the ocean death lasted, not more.

But what if we killed all the plants on the land as well as in the sea, so that no green plant could produce oxygen? We should, of course, starve to death, but, as a matter of academic interest, what would this do to our oxygen supplies? Let us, for good measure, assume that we drained all the world's swamps and marshes too, so that bacteria could no longer produce oxygen from nitrates. We would have destroyed all known mechanisms for producing oxygen except the breaking-down of water in the upper atmosphere. Let us assume that upper atmosphere physics in fact is not important, so that we can paint the worst possible picture of a world without our contemporary oxygen pump. Then, with all plants dead, and all the oxygen-producing bacteria denied a home, what would happen to our air?

Nothing very much, at least in the short run of a few centuries or so. This is because the total output of all the biological pumps is very small compared with the great mass of oxygen that has accumulated in the air throughout the grand marches of geologic time. The geochemist Wallace Broeker put figures to this observation very neatly by calculating the oxygen reservoir and the oxygen pump over a typical square meter of the surface of our planet. He imagines an invisible column, a meter wide at the base and square in cross-section, which stands on the earth and goes all the way up through the atmosphere until it reaches the vacuum of outer space. This column would hold 60,000 moles of oxygen. Since the atmosphere is well-mixed, his column would contain these 60,000 moles wherever it was placed, on a continent or over the sea. The yearly production of oxygen at the bottom of this column would vary from place to place, there being much more if the column sits on the land than if poised over the sea. But the average production at the bottom would be just 8 moles in a year. 60,000 moles in the reservoir; 8 moles produced annually. Clearly the production would have to stop for a great many years before anybody would notice anything.

Even if we waited a very long time, it is unlikely that killing all our plants and ending life on earth would ever produce any detectable difference in the atmosphere because this would stop the losses of oxygen from the air as well as the replacements. All our corpses would rot away, using oxygen to rot with, it is true, but it would take no more than a few tens of moles of oxygen per square meter to burn us all up to carbon dioxide gas; people, plants, bacteria, animals, soil humus, the lot. Then that unique mixture of oxygen and nitrogen would go on slopping about over the killed-off earth, while life was quietly started all over again.

Those self-styled spokesmen for the ecological profession who warn people that industrial life threatens to destroy the atmosphere spread the gospel of nonsense. It would be better for us all if they ponder the fable of the little boy who cried "Wolf, wolf."

We come then to the provocative conclusion that, although air is regulated in the very long term by the doings of living things, and might even have been made by ancient ecosystems in the first place, the accumulated store of gas is now so vast as to be almost independent of life processes for its maintenance. Yet we need *all* life processes for this comfortable conclusion to be safe, not just some of them. If, for instance, only the nitrogen-fixing bacteria were at work without the bacteria of swamps, the specter of mountains of nitrate rock can perhaps be raised, though only for the very long haul of the geologic time scale. Failing some selective killing of this sort, which would be very hard to do, there is little scope for folly in the handling of our oxygen reserve. The original creation of air out of oxygen and nitrogen has been done on too grand a scale for any process we know of, either living or dead, to make much difference.

There is, however, one aspect of the air that is regulated, and in a very subtle and delicate way. This is the carbon dioxide supply. Carbon dioxide is present in the atmosphere at an average concentration of about 0.03 percent, an amount small enough to be hard to measure accurately without the tools of a modern analytic laboratory. This tiny concentration is maintained by an elaborate array of interacting mechanisms, rather as the nutrient supply of the oceans is kept dilute and constant over geologic time.

There are 50 atmospheres worth of carbon dioxide in solution in the world oceans. The gas is readily soluble in water, thus the oceans are generally saturated with it. The chemistry of its solution equilibria is not the simplest, since it involves carbonic acid, carbonates, and bicarbonates, but this does not alter the outline of the story. There is 50 times as much carbon dioxide in solution in the sea as there is in the air and, since the seas and the atmosphere are touching one another, a free exchange takes place between the two.

The 50 atmospheres of carbon dioxide in the sea act as an enormous shock absorber for the air. If the air was depleted of carbon dioxide, the sea would lose gas; if something happens to give the air too much, the sea would soak up the excess. The 0.03 percent of carbon dioxide in the air is thus the result of a chemical equilibrium between the air and the sea that should be very hard to upset. Nor is this all, because the shock absorber is itself shock absorbed.

The sea collects bicarbonates and carbonates not only by absorbing carbon dioxide from the air but also from rivers after rains on the land endlessly wash away limestone rocks. The sea deposits the excess, as it does with all the other dissolved things that come to it. We know the carbonate deposits of the sea well; they are the coral reefs and the white calcareous oozes that will one day be mountains of chalk like the Dover cliffs. There is an endless chemical cycling of carbonates like that of the nutrients described in the last chapter. Cycling keeps the concentration of dissolved carbon dioxide at 50 atmospheres' worth, acting as a shock absorber to the ocean reservoir of carbonate. So the concentration of carbon dioxide in the air is kept constant and rare by a solution equilibrium between air and sea, and the concentration in the sea is itself kept constant by the process of depositing excess carbonate on the sea's bottom.

All this is passive chemistry, an inanimate system that regulates the carbon dioxide in the air and determines the conditions under which life must be lived. As with the oxygen and nitrogen cycles, living things are involved at some stages in the system, particularly as animals and plants work to build coral reefs or deposit tiny calcareous skeletons in the bottom ooze. But it is likely that living things are not really necessary to the process because much of the carbonate is deposited by simple chemical means. If corals did not build reefs, the carbonate would be dropped from the oceans anyway, though not perhaps in the same place. Ecosystems do not regulate the carbon dioxide supply. Physical systems do this, and the living things of ecosystems make do with what they are given.

So I conclude that the regulation of the air is largely done by chemical and physical processes at the surface of an earth flooded with light. All life adapts to what these processes give it. In particular, the follies of people cannot seriously endanger their air supplies.

Yet there is one caveat, and it concerns the carbon dioxide supply. We now use petroleum and coal for fuel. When we burn it, we add fresh carbon dioxide to the air. This carbon dioxide was in the air in the first place, of course, millions of years ago, but the plants that made sugar out of it died in places where they could not rot and so were kept as coal or oil. At the same time this loss to the air was made good by the ocean shock absorber and the morsel was "forgotten" by the system. But now when we burn it, that ancient carbon dioxide comes as something new, a fresh influx of gas to the air. The coal and oil reserves that we expect to use up in one or two centuries represent several atmospheres' worth of carbon dioxide, so we are doing our best to multiply the concentration of the carbon dioxide of our air by several times.

When geochemists first pondered the implications of this burning of fossil fuels they saw nothing to be alarmed about because of the complex system of shock absorbers. There was, and is, no doubt that the oceans can cope. An extra few atmospheres' worth of carbon dioxide is nothing to the 50 atmospheres' worth already in the ocean, or the 40 thousand atmospheres' worth that are locked up in limestone mountains and coral reefs. All the carbon dioxide from our burning of fossil fuels will pass into the sea and thence to its mud. The concentration in the air when the fuels are spent and our civilization is no more will be nicely adjusted back to 0.03 percent by volume. This is certain. But some recent measurements of carbon dioxide in the air by geochemists have been surprising all the same.

In Antarctica and on a high mountain in Hawaii records have been kept of the concentration of carbon dioxide for several years. These two places were chosen because they were far from heavy industry and were windy places. They should be free from local effects like the belching smokes from factory chimneys, and their air could be treated as typical average global air. And in both places the carbon dioxide has been rising, steadily, year after year. We had placed our faith in the shock absorbers a little too soon.

What has gone wrong with the shock absorbers then? Simply that it takes time for the oceans to soak up the excess. There is no doubt that a new equilibrium is quickly set up between the surface of the sea and the air, but then the oceans have to be stirred up so that water with the extra dissolved carbon dioxide can get to the bottom and be replaced by other water to take its turn in the blotting-paper queue. The oceans are five miles deep and are stirred by the winds only slowly. It takes a hundred years, in fact, for the oceans to turn over once.

The shock absorber is working, but slowly. In the end it will soak up all the polluting carbon dioxide from our fossil fuels and pass them to its mud. But meanwhile the shock absorber has more than it can handle, and there is a temporary pile-up of the gas in the atmosphere. The most likely outcome is that the concentration in the atmosphere will nearly double before the oceans begin to catch up and bring carbon dioxide back down to its old proportions.

The carbon dioxide in our air will increase, thus, from about 0.03 percent by volume to about 0.06 percent by volume. This will not be catastrophic for life. On the other hand it will certainly make a difference. The worry is that we do not know what all the effects might be. We are embarked on the most colossal ecological experiment of all time; doubling the concentration in the atmosphere of an entire planet of one of its most important gases; and we really have little idea of what might happen.

# Chapter Eleven. The Curious Incident of the Lake in the Now Time

"Is there any point to which you would wish to draw my attention?"

"To the curious incident of the dog in the night time."

"The dog did nothing in the night time."

"That was the curious incident," remarked Sherlock Holmes.

From *Silver Blaze* by Sir Arthur Conan Doyle

THE lake a poet might like, a cool clear lake, blue, with mysterious depths, probably in the Alps or other mountainous place far from industrial cities, will not be fertile. It may occupy a raw hole left by a mountain glacier, or it might be surrounded by alpine meadows from which very little nutrient trickles. The lovely blue of its waters is the blue of the oceans, being the glow of light shining through clear water in which very little grows. Blueness of water, coolness, and infertility go together, an idea very simple and easy to understand. But scientists do not like their simple observations to be understood by others. So they call infertile lakes "oligotrophic," which means the same thing in Greek.

This beautiful, useless, oligotrophic lake will certainly freeze over in winter, when its water will be cold all the way to the bottom. In the spring the ice melts and the winds blow the waters. The lake is stirred by these winds right down to its bottom. This means that the cold, infertile water of the lake is loaded with oxygen taken from the wind.

With the summer calm, the currents and the waves depart. The sun heats the surface. Warm water expands, it gets lighter, it floats. Soon there is a thick layer of warmed water at the top, which is floating on cold bottom water like an oil-spill on a puddle. The lake is now layered; warm on top but cold underneath. And these layers cannot be mixed together by winds of ordinary strength. An attempt by the wind to force the warm top water down to the cold depths would be like a swimmer trying to sink a big rubber ball in a swimming pool. Anyone who tries this knows that it cannot be done, because the resistance of the ball to sinking is obstinate. In a like manner the warm water at the surface floats obstinately and cannot be made to sink.

From the point of view of anything living there, the lake has now become two, quite separate, lakes. There is a top lake, which is warm and which sloshes about under the air. And there is a bottom lake, which is cold, motionless, and completely cut off from the air by an invisible, floating roof of warm water.

All animals, even aquatic ones, need oxygen, and the prime source of oxygen is still the air. Being cut off from the free oxygen of the air can have serious consequences, and a critic of some of nature's plans might well say that choosing to live in the deep water of a lake has obvious hazards. Yet a few animals get away with the niche of deep-laking in oligotrophic lakes. A lake trout does it for one. It does come to the surface to hunt, but it spends much time in the depths. It gets away with this hazardous undertaking as an indirect result of that remarkable infertility that lets the water glow blue.

If the water is infertile, very little can grow in it. An infertile lake is like the open sea, with a very thin population of planktonic plants and a truly tiny mass of animals based on them. Since there are very few animals, there is very little call on the oxygen reserve. A few trout in a big lake can probably get through the summer with the oxygen left in the water from the spring mixing, though one might guess that a matriarch of a trout might feel relief when the cold air and winds of October puncture that balloon of warm water on the top and mix the lake once more. But, if the lake is of the prettiest blue, and therefore very infertile and almost useless, the few trout who live there may actually find that they have a private oxygen supply outside the reservoir held by the water.

When a lake is so infertile that it glows blue, sunlight reaches right to the bottom and in this lake the minute plants of the plankton can live deep down in the cold water. Some of them even live lying on the bottom. It is so cold down there, and there are so few nutrients, that the plants do not produce much. But their factories tick over all the same, and as they make their sugars, they release a trickle of oxygen to the water. And so "trouting," and other professions of liking-to-be-where-it-is-cold, can continue with secure oxygen supplies.

Yet many lakes, even of the temperate north, are fertile. They are richly endowed with nutrients that have poured into them from good soils in the surrounding drainage. They are turbid green and brown with floating and swimming life. They are never blue. They are splendidly productive, perhaps producing more food per acre than the finest farmland. They are like the ponds made by monks of medieval monasteries so that the brothers should not lack for fish on Fridays. Their appearance, as well as their usefulness, is an obvious and simple result of their being well-fertilized. So we of the profession must hide this profound truth from lesser beings by calling these lakes "eutrophic," which is Greek for "fertile."

A eutrophic lake may freeze over in winter like any other, and the life in it will shut down its factories while awaiting better times. It too will be stirred during the windy days of spring, so that all its water is cold and filled with oxygen, which it soaks up when it takes its turn in the breaking waves. Then it will settle into layers under the calm summer sun so that a cold bottom lake is cut off from the air by an upper, floating lake of warm water. But the consequence of this layering will be dramatically different.

In the warm, fertile water of the upper lake, life will bloom abundantly. The tiny plants of the phytoplankton will crowd together at the top, climbing over each other for the light as do the striving trees in a thick forest. If the eutrophic lake is very fertile, a bloom of blue-green algae may float on top of the water. And this bloom of life is thick enough to completely shut out the sun from the depths below.

It is dark near the bottom of a eutrophic lake in summer, pitch dark. No plants can live down there. This means that no oxygen can be got from bottom algae as in the depths of an oligotrophic lake. The only oxygen available is the reserve taken down in the spring. On the face of it, this need not matter very much because the water remains cold and a trout or two might be able to make do with this oxygen reserve until merciful October came, perhaps even having the benefit of occasional hunting trips in the life-zone overhead. But the trout, and other cold water animals, are not allowed to keep the oxygen for themselves. From the dark shadows overhead there is falling all the time a brown snow of rotting corpses.

Where there is much life there is much death. Near the surface of a eutrophic lake there is much life, but the living community does not have the tiresome duty of taking care of its many corpses. It exports them, by gravity, to the cold, suppressed lake below. They rot down there, and the bacteria that eat them use large amounts of oxygen in the doing of it. They take this oxygen from the reserve. If the lake is really fertile, the bacteria are likely to use all the oxygen in the water before the summer is over. Anything else that wants oxygen, like trout, has to die.

We come now to a curious contradiction of our times; people prefer fertile, productive fields, but they like their lakes to be infertile and unproductive. Indeed, they often defend the merits of infertile lakes with passion, branding he who puts phosphates or manure into water as the archetypical polluter (literally the maker of dirtiness). They borrow scientific jargon and make it worse, accusing the man with the fertilizer of the high crime of "eutrophication." Yet he who puts phosphates and manure on farm fields is doing a public good and is a pillar of rectitude in his community. Eutrophicating fields is "good"; eutrophicating lakes is "bad."

People like their lakes unproductive partly because blue is pretty and catching trout is fun, but also because rich, fertile lakes tend to stink. A stink is a well-known property of a corpse, and the rich life of a fertile lake is bound to produce abundant corpses, which will make the water smell a bit. If one has a cottage on the lake, then an aversion to smells is understandable. If, however, one lives in a monastery or a Chinese commune then a smell is a small price to pay for a good yield of carp, and in those communities one can acquire a reputation for public rectitude by shoveling in the manure. Even in these places you use non-human manure, for reasons of hygiene, but this is not part of the lake-fertility argument.

But polluting a lake with manure and phosphates has come to mean more than just killing trout, growing algae, or making a smell. The idea is abroad that one is doing something bad to the ecosystem itself, that definite, irreparable harm has been done by fertilizing a lake. A news magazine once had a headline of "Who Killed Lake Erie," and the writer thought he meant more than "Who killed a few trout." Students of lake systems themselves are responsible for the talk about killing lakes. Let us see what they mean by it.

Think again of the clear blue waters of a typical oligotrophic lake, where there is oxygen in the deep water all summer long. Even the mud on the bottom of this lake is supplied with oxygen and is usually a reddish brown color. The red speaks of rust-like oxides of iron. All the other active minerals in the mud in contact with the water will be likewise well-oxidized. And oxidized minerals in mud have the property of hanging on to things such as phosphates, potassium, and nitrates. When nutrients enter the water of an oligotrophic lake, therefore, many of them are seized by the mud and held in escrow. It is a system like that which keeps the blue oceans infertile forever, however many nutrients the rivers bring down to the sea. So an oligotrophic lake has a chemical system that keeps it infertile. This is why a lake founded by a glacier more than ten thousand years ago may still have blue water in it.

Now think of the state of an eutrophic lake in summer. This lake has no oxygen left in the bottom water because the bacteria have taken it all away as they worked-over the corpses. The surface mud, too, will have lost its oxygen as a consequence, and will be gray and smell of hydrogen sulphide if stirred with a stick. It happens that mud like this is much more active chemically than oxidized mud, and it has the particular property of making nutrients soluble. Nutrients keep landing on the mud of an eutrophic lake, because they come down with the corpses, but the mud makes them soluble and sends them back into the water. Once a lake has become fertile, it tends to stay that way.

There are, therefore, quite different chemical systems operating in fertile and infertile lakes. In each lake the system works to maintain the status quo. We may legitimately worry that brutally dragging an infertile lake across the great chemical divide with a massive dose of fertilizer is doing something we cannot reverse and may live to regret. This perhaps makes clear some of the worries of the defenders of lakes, but it still has not brought in the "death of the lake" motif. To do this we need to think of what the passing years do to the basin of a lake.

Every year a lake gets shallower. This is the fate of all lakes just as mortality is the common burden of mankind. Mud collects in their basins year after year, and will do so until the hole is full and the water totally excluded. The typical glacial lakes of North America and Europe are already half-gone by this process and will vanish in another fifteen thousand years, if another ice age does not come to scoop them out again. The aging of a lake is a process of filling it with mud, and the killing of a lake is the filling it with mud utterly.

But it is the life and the ecosystem of the deep water that feels the aging process first. In summer the surface water is always warmed through to the same thickness, separates itself from the bottom water, and floats over it, so that the passage of twenty thousand years may make little difference to the size of this upper lake. It is the cold, lower lake that grows smaller each year. As it gets smaller, so the size of its oxygen reserve decreases. Inevitably a time will come when the oxygen reserve of the shrinking lower lake is so reduced that it will not last the summer through, even in a lake that has been oligotrophic since before the dawn of history. And in that critical year the oxygen will go out of the bottom water for the first time, the mud will lose its oxygen for the first time, and nutrients will be released from the mud back to the water in quantity for the first time.

One of the results of this process is, therefore, that an infertile lake starts extracting nutrients from the store in its mud in its old age and collecting them in its water. Instead of a chemical system that keeps it unproductive, it acquires a system which promotes fertility. In the ugly jargon of students of lakes, "an oligotrophic lake can become eutrophic by a process of infilling."

This is why some writers call polluting a lake with fertilizer, "artificial aging." They say that the lake will have become fertile in its old age eventually if we had let it alone, and that we forced it into the old-age state by making it fertile early on. But it should be clear how wrong-headed those writers are. Natural aging means filling up the basin with mud, and amongst the consequences of doing this is that the water becomes more fertile at the last. It does not follow that putting fertilizer in is tantamount to filling the lake with mud!

Lake Erie is neither dead nor even aging very fast. Nor are any other lakes that have been afflicted with phosphates, sewage, and garbage. They are merely very fertile now when once they were not. They are not dying. They are crawling with life. This is what is wrong with them.

Yet, if polluting a lake with fertilizer does not age or kill it, it does change things drastically and we have already remarked that lakes tend to keep to whatever level of fertility they have reached. Once we have made a lake fertile a critic can perhaps point his finger and say, "Now you have done it." But even this is not true. The system that keeps a fertile lake eutrophic needs a lot of help from outside.

All lakes, particularly fertile ones, bury the corpses of

their inhabitants in their mud. Much of the corpse matter, of course, is rotted down by bacteria and released back to the water, but there is always an excess that gets buried. With these buried corpses go nutrient chemicals. The mineral sediment, too, drags nutrients down to the bottom and, even in the smelly mud under water without oxygen, hangs on to some of them until they are well and truly buried with no possibility of being redissolved. No eutrophic lake will remain fertile entirely of its own accord because it must receive as many nutrients from outside as it buries in its mud. So, if we make a lake fertile by polluting it, we must go on polluting it in order for it to stay fertile. If we stop, the lake will throw off the symptoms of fertility.

Lakes are self-cleaning systems. If we stop dumping sewage, fertilizers, detergents, and garbage into any polluted lake, its waters will come clean again, by themselves, without any benefits of technology, in a very few years or decades depending on its size and the flow of water. The only way to kill a lake is to fill it in and pave it over.

In a century or two, when we have run out of fossil fuels and phosphate rock, all our polluted lakes will revert to their former states. If we now keep some lakes infertile, so that animals that like cold water with little to eat in it do not become extinct, the temporary condition of the remainder is one of our lesser worries.

# Chapter Twelve. The Succession Affair

"As an organism the formation arises, grows, matures, and dies." (. . . . .) "Furthermore, each climax formation is able to reproduce itself, repeating with essential fidelity the stages of its development. The life-history of a formation is a complex but definite process, comparable in its chief features with the life-history of an individual plant."

From "Plant Succession: An analysis of the development of vegetation" (1916) by F. E. Clements

IF the planners really get hold of us so that they can stamp out all individual liberty and do what they like with our land, they might decide that whole counties full of inferior farms should be put back into forest. And if they had a strong enough police force and plenty of weapons, they could do it. They would plough the fields where the last crops had been, plant their seedling trees in depressing rows, hoe between them, put down fertilizers, and spray assorted poisonous chemicals. And the forest would come up. It would do so in the single lifetime of a tree.

Nature can put forest back onto farmland too, without benefit of hoe or chemicals. But it usually does the job slowly. Nature gropes toward forest in roundabout ways, through weeds and shrubberies to a changing pattern of trees. When farms on land that was once forest are abandoned, the fields first become choked with weeds. In two years or so a different and tougher set of weedy plants takes over. We have a knee-high meadow of Michaelmas daisies, or golden rod or coarse pasture if cattle or horses are allowed to roam in it. The fields look like this for a few years, but then begin to acquire the rough look of real wasteland, with briar patches, thorny bushes, or, in some places, small pine trees. This is the typical sort of regenerating wilderness found on the outskirts of many a large Western city where the real-estate operators are holding onto land until the price is right for them to build houses. Only a decade or two might be needed for the fields to close over into dense bushland. But usually it will be a long time before young trees of the real forest begin to penetrate the bushy growth.

An abandoned farm can be covered with woods of a sort in the lifetime of a person, but it is likely to be a watcher's old age before the first trees of the original wild forest will make their appearance. The weeds and choking thorn-bushes have been an overture to the growth of the forest. And yet potential parents for the forest trees may have been lining the field all the time in its hedgerows, or even scattered across it for shade in which ploughmen and their horses once rested. But rarely will a baby forest grow until the weedy preliminaries are done with. Foresters could have planted the forest trees in the first year, but nature provides a succession of formal occupations of the field by other plants before she is ready for the trees.

This progressive occupation of abandoned fields by a succession of different plant communities has been often observed. It always happens. First come in the annual weeds, plants whose niches include the ability to have tiny seeds spread far and wide on the off-chance there will be a patch of bare earth in which they can grow for a season and scatter some more seeds. Next come the perennials, herbs that bite into the ground with resistant root systems and hold onto the fields year after year— then the bushes, then the scrub trees of woodlands.

This much is fact. That the process will end with the coming of the primeval forest is conjecture because scientific man has not been around long enough to see the whole process through. It is, however, a rather safe conjecture. We can see the forest trees coming in and we can look at ancient forests disturbed in olden times and see how the mix of trees changes ever more closely toward the original virgin growth. The whole, complicated proceeding of a succession of plant communities always occurs when disturbed land is let alone. It is so regular, predictable, and orderly a process that a good local botanist can tell you the date a farmer quit farming merely by looking at the plants that are now growing on his land.

But now comes fancy. Once upon a time the virgin forest held sway over all this land. Then came the settlers with their axes, their teams of horses, and their ploughs. They cut into the virgin forest, and when they had finished there was an open land of crops. Yet hovering near it in woodlots and hedgerows, the remnants of the plant societies of the virgin forest hung on. Every year the open fields were assaulted by small plants, the pioneers of the forest, which the farmers called weeds. The farmers drove the weeds back, though with much labor. At last the farmers tired of the struggle and went away. The weedy pioneers of the forest grasped the land firmly, making it possible for the consolidation forces of the perennial herbs to take over. The forest then continued building with shrubs and scrub trees, until, in the fullness of time, the virgin forest was repaired. The living vegetation has repaired its wounds even as individual plants or people heal theirs.

When botanists first looked at vegetation with an ecologist's eye, they saw those nation states of trees and the plant societies of which I wrote earlier, but they also discerned the grand marches of the ecological successions; orderly, predictable sucessions that would reach a climax in the plant nation or society proper to the place. Plant succession was a property of a whole community in which each individual kind of plant had a job assigned to it. Here, then, is one more piece of evidence for those who look for grand designs in nature.

If one went out looking for succession, one could find it in any country and in any place where vegetation was growing over the bare spots. The wounds caused by hurricanes and fires are cauterized and repaired by the same successions that reclaim a farmer's fields. There are successions in tundras and prairies as well as in forests. There are even successions on land that never held the climax vegetation before, such as the land made from filled-in waterways, or rocky scree on which thin soil has collected. The actual species of plants involved in all these succession changes from place to place but in any one country or terrain the species are always the same. And everywhere the order in which they appeared was fixed; first the short-lived or annual weeds, then tougher perennials, and thence to the climax formation, through woody shrubs if the climax were forest, directly to the perpetual herbs if the climax was grassland.

It was at once apparent that the plants made changes in their physical habitats. The annual plants always pioneered where there was bare, ploughed or scorched earth, naked scree or fresh mudflat; but the pioneers cloaked this surface with their foliage and afterward with their decomposing parts. The soil was better when they left it to the perennials who came after. The perennials built up the soil still more, and they built up the soil nutrients too, as chemicals collected in their litter and as the nitrogen-fixing bacteria who lived with them left a legacy of nitrates. Thus the habitat in which the shrubs make their start is more congenial than the barrenness faced by the pioneers. And it is made better still before the climax trees come.

This discovery made it possible to see one practical purpose in this process of succession. The vegetation prepared the site. The elaborate business of a succession of plant communities was necessary to make the land suitable for the climax plants. This was the reason that nature went her roundabout way in replacing a forest. Whole communities of plants were working together to make the land fit to live in.

None of these ideas is new to modern science; the facts of succession were seen, and noted, by countrymen long ago. Some say that Aristotle was the first to write on the theme, though this probably reflects the fact that not many writings from earlier civilizations than that of Greece have survived. But succession began to be exciting only seventy or eighty years ago as the first unifying theme in ecology. In Russia, Denmark, France, and England botanists were writing of this remarkable orderly process while Queen Victoria was still alive. And in the wilderness of the American West, a young, articulate man, more daring than the European savants, developed a general philosophy on the workings of nature from what he saw the plants do, the impact of which is with us still.

Frederick Clements was born in the prairies in 1870, just after the first settlement. The last buffalo herds were exterminated while he was a youth. He was seven years old when Custer rode to his last futility on the banks of the Little Bighorn River. He took one of the first doctoral degrees of the new University of Nebraska at the age of 23, and for a thesis called "The Phytogeography of Nebraska." At a time when the first European ecologists were beginning academic arguments about the definitions of terms, the young Clements took a mule train over a wilderness state the size of a European kingdom and described it in his Ph.D. dissertation.

What Clements saw from his mule train was the virgin prairie starting to heal the wounds made by men and animals. It did so by succession, sending out pioneer plants into the ruts made by the wheels of wagons and across the trails worn by the vanished herds of buffalo, consolidating the land thus won with perennial herbs, and building toward the climax formation of prairie grasses. In a like manner the prairie crept its way across new land, feeling across sand dunes and puddles with its pioneers, covering the bare spots, preparing the soil for the climax grasses. "As an organism," the whole formation of plants was arising and growing to maturity.

Clements wrote well of what he saw. He convinced many botanists of his generation and the next that whole plant communities had individual lives. We must not think only of separate plants or separate species, because these were but working parts of a much larger natural entity, the plant community or formation. His was the first passionate statement of the existence of natural communal entities that must be regarded as whole things in their own right. Three-quarters of a century later we still echo Clements's passion when we talk of being careful in tampering with ecosystems lest we damage the working of the whole thing and commit "ecocide." All this came about because a young man saw a prairie wilderness regenerating by succession in his own lifetime, and because he thought about what he saw.

But is it really so? Is there some mystic organization

beyond the species level that fits communities together? Or could everything that Clements saw be explained more simply by an hypothesis that lets each species in the plant community act selfishly in the pursuit of its own Darwinian fitness? The Darwinian idea of every species struggling with every other is not so emotionally pleasing as the Clementsian dream of species cooperating for the good of the community, but it is simpler.

The facts of succession are at their most impressive in forests. Short-lived annual weeds give way to perennial weeds that live on year after year. Tough briars and bushes climb over these perennial herbs until they are themselves climbed over and shaded out by trees. As this goes on, the species richness of the community grows, for although there may be only two or three kinds of annual weeds in the first year, dozens of species will be growing in the scrubby woodland fifty years later. As the species richness grows, the soil too is made richer. Any theory of succession must explain all these events. But when we began to look closely at the process we found two things that were decidedly odd.

The climax formation often contains less species than the communities of earlier succession stages. This is an easily demonstrated fact. Any mature forest of North America or Europe will reveal a certain monotony. A few kinds of trees dominate the forest and to these may be added the bushes and creepers of the understories together with the flowers of the forest floor. But scrub woodland with open spaces may contain dozens of kinds of trees and shrubs, as well as a wealth of herbs between them. It seems that the climax of Clements's vision discards many of the members of its community once they have served its purpose, which is perhaps not so satisfying to the emotions as the rest of his message.

Second, we find that the climax is not only poor in

species but is inefficient as well. Should one measure the
efficiency with which a field of pioneer weeds converts
solar energy into the calories of sugar one will find that
they do very well, quite like Transeau's cornfield of
Chapter Four, in fact. The perennial herbs that follow,
the pioneers and the various communities of shrubs that
follow them, do just about as well as pioneers and corn.
This need not surprise us, because we have come to expect that all plants will be equally efficient because all
are restricted by the common shortage of carbon dioxide. But some measurements have been made that lead
to the claim that the climax forest converts solar energy
with less efficiency than that of any of the herb communities which preceded it. This is a surprising claim.
We are told that the climax trees, those that have replaced all others in a ruthless Darwinian world and who
will hold the habitat against all comers indefinitely, are
less efficient than the plants they have replaced. It does
not seem right.

But let us take the facts of succession one by one and
see if reason cannot explain it all. We have to explain an
orderly progression of communities that improves the
soil as it goes, that this ends in a climax community dominated by a few kinds only, and the remarkable possibility that trees of this climax community may be inefficient
in the vital thermodynamic sense.

Annual weeds have niches that let them take advantage of some sudden opportunity for unrestricted
growth. They use the small-egg gambit, spreading very
large numbers of tiny seeds far and wide. When these
seeds germinate, they must have a lucky break if they
are to grow because the young plant will have almost no
reserves of food and must be on its own from the very
start. But in moist, bare ground, it can manage because
the sun will provide the calories it needs and it will not

have to fight established plants for space. So an annual weed is in the business of living where accident has cleared away the competition. We call it an "opportunist" species* because its strategy is to broadcast seeds everywhere so that in every part of the landscape some are waiting for a favorable accident. A chance for such a plant will always come in every habitat, even in virgin forest, for all it needs is the fresh earth of a rodent burrow, the collapsing bank of a stream, or a place where wind and fire have torn up the sward.

On bare earth the tiny weed plant grows quickly and without being molested by others. Its strategy must be to make seeds as fast as possible because the more permanent plants of the place will soon be pressing on it and the ensuing struggle for the sun would take the energies of the plant away from producing its mass of seeds. The opportunist plant does not waste effort on competing. It throws every calorie into seeds, and, when the competitors or physical adversities such as winter do close in, it surrenders and dies. This is a good strategy for life and very successful. Plants that we call "weeds" maintain it successfully even against the best efforts of farmers to frustrate them.

The obvious alternative strategy to that of the opportunist weed is to hold tight to what one has, to live for years, and to keep up a steady reproductive pressure all the while. Plants who adopt this strategy are called "equilibrium" species. Perennial herbs tend toward this strategy. They make big roots or underground storage organs that let them last the winter and thus begin at an advantage the following year. This store for the winter must be made at the cost of fewer seeds, because a plant

---

* "Opportunist" species or strategies are now often called "r" species (after symbols in growth equations) and equilibrium species or strategies are called "K" species.

can only do one thing with food calories. An annual plant spends all its coin on making seeds. A perennial spends some coin on its own food reserves as well and we suspect that it spends some more coin on holding off the competition. An equilibrium species plans for babies in the future rather than babies now. And clearly there are many possible compromises between the extremes of living forever, on the one hand, and killing oneself having all one's babies now on the other.

Forest trees do not start growing on the fields when the farmer quits because they are equilibrium plants that have not gone into the business of soaking the habitat with tiny seeds. The weeds come in first because they have. This is the prime reason that nature goes about replacing forest in such a roundabout way.

Succession will begin with weeds on an abandoned field just as it will on any bald patch made by nature. The opportunist weeds have it staked out with their seeds in advance and will achieve a quick crop. But that is all they can do because the next spring the herbs with equilibrium habits bound up from the hidden roots, offer overwhelming competition to any tiny plants that pop out of seeds beside them, and inherit the field. Many of the perennial weeds are daisies and their kind, plants that scatter tiny seeds but that also make magazines of food to tide them over winter. They are plants that have hedged their bets, half-opportunist and half-equilibrium. They fall in turn to plants who have done less hedging.

Briars and shrubs invest still more of the coin of calories into resilience. They make woody parts that let them climb over the herbs and steal the sun. These woody parts cost not only calories to make but also many more calories to maintain, and these calories must have been taken from seed production. So the shrubs are late arriving but ferociously effective when they do come.

Then come the trees, which upstage even the shrubs. Clearly the process will go on until it no longer pays to divert calories from seeds to making giant bodies. What we have then is the great tree of the climax forest.

Many of the facts of succession are now explained. Nature goes her roundabout way in the building of a forest because any community contains a mixture of opportunist and equilibrium species. The opportunists get there first and are replaced by plants that show successive compromises with the equilibrium strategy. All the orderliness and predictability of succession is explained by this. Even the upgrading of the soil is explained, for it is no more than the inevitable consequence of having plants grow on what was once bare ground. Upgrading soil is more an effect of succession than a cause.

Our understanding of succession is at the stage where we can say that there is nothing mysterious about the predictable replacing of one vegetation type by another. The apparent community properties are simply explained as the consequence of numbers of different Darwinian species each doing its own thing. The fast movers come in before the slow. But there remains the mystery of the dominant few in the climax forest. Why is it, for instance, that nearly all the trees in an American beech-maple forest are beeches and maples in spite of the fact that there are 18 other species of tree for them to lord it over? And then there is that strange matter of the inefficiency of some climax trees.

Ecologists still argue hotly about dominance, but an explanation that is at least partially satisfying has recently appeared. The crucial observation for this was that there does not seem to be any dominance in some lowland tropical rain forests. In the flood plain of the Amazon River a hundred species of trees may exist in an acre, with none of them so much more common than the rest that they can be called "dominant." This very diverse forest with no dominance clearly contrasts with temperate forests where there are only a few species of tree, dominated by one or two kinds that are superabundant. We might never have become so interested in the problem of dominance were it not for the fact that most of the world's universities, and the ecologists they shelter, are in the temperate belt.

Some tropical forests do exhibit dominance (the widespread *Mora* woodlands, for instance), but it may be that these live in places where seasons of different kinds give their tropical setting much that is in common with the habitats of temperate forests. It seems, however, to be true that, in the wet tropic places without seasons, the forests are both very rich in species and without dominance. When a botanist, Daniel Janzen, turned a very open mind to the way the trees of these forests were scattered about he came upon what may well be a decisive factor to account for dominance—insects.

Many kinds of insect are in the business of hunting seeds and are very good at it. Most tree fruits are probably attacked by boring insects, or have attached to them the time-bomb of an insect egg, before they are dropped from the parent tree. Other insects wait for the fruit on the ground. Squirrels and mice join the seed hunt. In a tropical place without noticeable seasons, many of the insects and rodents manage to keep large populations year round, year after year, with the result that the attack on the seeds of trees is very strong indeed. Furthermore the trees act as beacons to the attackers, so that the appropriate seed-eating insects congregate there, and rodents can home in on a fruiting tree from considerable distances. Janzen was able to show, by scattering seeds himself under trees, that there was a seed death-zone under a tropical forest tree in which the total attack on seeds was always 100 percent successful. It is impossible for a baby tree of the tropical forest to get started under or near to its mother. Only seeds that are carried by parrots or other dispersing animals beyond the circle of waiting enemies have any chance of germination and survival. This means that each species of tree of the tropical forest lives spaced out, the gaps being filled by other kinds of tree with different seed predators or trees that fruit at different times. Thus, dominance of the many kinds by the few cannot occur in this kind of tropical forest.

Such were Janzen's conclusions about the spread of insects and trees in the forests of Costa Rica. They provided an answer to the question how dominance can occur in northern forests where there are winters, and in the tropical forests that are subject to climatic seasons. Where there are seasons, and where the weather fluctuates within wide limits, the sizes of insect populations vary greatly. Insect numbers fluctuate with the weather. When the insect seed-hunters of a northern forest have had a good year (from the insects' point of view) their attacks on seeds may be as devastatingly efficient as those of their cousins in the lowland tropics. But in a year when the insect numbers are low, their attack is muted, and most of the seeds escape. A dense crop of baby trees can then get established and starts to grow underneath their mothers. This leads, a generation later, to forest-stands of only one or two species; what we call dominance. These few species can go on making babies indefinitely, because the chances are good that in each tree generation there will be years when the insects go away. Other species never get a chance, and the climax forest casts out all the partly opportunistic plants that grew before. This explanation leaves dominance as one more consequence of Darwinian self-interest. The presence of a few ruling trees in the climax that Clements saw need have nothing to do with self-organization by plants. The general ineffectiveness of foraging insects in the northern places where Clements and his kind worked is a more plausible cause.

Janzen's insight into the insects that mold the climax forest is 1970 ecology; it took fifty-four years after Clements published his great book before we could offer at least a partial explanation to the observation he publicized so eloquently. The explanation for dominance is still only partially satisfactory because there have been few tests of it in places other than Janzen's Costa Rica, and many botanists are not convinced that insects usually have the effect Janzen claims. But it is a satisfactory explanation as far as it goes. The matter of the productivity of trees of the climax forest is the subject of a debate even more recent still.

The claim is this: in the climax community productivity is less than in earlier succession stages, which implies that the efficiency of energy conversion is less in the climax. The evidence to support this surprising claim is of two kinds, some from production measurements on forests but some, and this the stronger, from measurements on a quite different ecosystem; a microcosm in a laboratory. Because much argument has been based on the results of looking at laboratory microcosms it helps to listen to the tale they tell before getting back to the forests.

These microcosms are bottles of water with algae inside and an artificial light shining on them. The start of a microcosm is water, a mineral solution, and starters of various kinds of algae. There is then a succession of population events in many ways analogous to the land plant successions on old fields. The pioneer plants of this succession are microscopic planktonic algae, and the populations of various kinds of these ring the changes in the open water of the bottles rapidly, in ways that are directly comparable to the successive invasions in the early years of old field successions. These changes climax with growth being transferred from the open water to the sides of the bottles and taking the form of a thick brownish growth of filamentous algae obscuring the glass. When this happens, there are very few microscopic plants left growing in the open water, which looks clear. This sequence can always be predicted in laboratory bottles deliberately left alone under a light but it is also familiar to anyone who has neglected a fish aquarium. In the stagnant days when it has been forgotten, the glass sides of an aquarium are covered with algal scum but the open water looks remarkably clear. Since things do not change very much from this state it is legitimate to say that a climax has been reached.

It is easy to measure plant production in one of these microcosms by monitoring the chemistry of the water, and we have very good laboratory data to show that productivity declines as the succession proceeds. Production is high as long as the plant life is in the open water, but when the growth on the sides of the climax is established, the rate of production falls and remains constant at a level well below that of the early pioneer stages. It follows that the efficiency of the climax in this succession is less than that of the communities that have gone before. This is really so. When similar claims are made for the climax forest, the prior experience of the microcosms can prejudice the result.

But the low productivity of the climax in bottles of water is easily explained. Many of the nutrients originally present in the water have been taken out of solution and stored in the bodies of the attached algae. Many of these algae are moribund, and the nutrients held in their senile bodies are not being efficiently used. These nutrients limit productivity in aquatic systems, and their effective removal from the system also lowers productivity of the open water community. Hence the low productivity of the climax is caused by the sequestration of nutrients. But this circumstance has no direct bearing on the question of productivity of climax forests.

The direct evidence in support of the claim for low productivity in climax forests comes from the experience of foresters that climax trees do not grow very well, and this experience is backed up by many direct measurements of production of different stands of trees, particularly by Japanese foresters working in Southeast Asia. This experience, and these measurements, have led to a widespread belief amongst ecologists that production by trees of the climax forest is usually less than that of trees of the successional stages that precede the climax. We have thus found ourselves in the position of needing to explain how it can be that a tree that inherits the forest is less efficient than the trees that it replaces. It seemed to some of us that the answer must lie in the adaptations of a climax tree to a youth in the shade of its parents. This must surely put engineering restraints on tree design because a design that works well in the shade of the forest floor might not be so good when exposed to full sunlight in the canopy many years later. I developed this argument in my own textbook, published in 1973, but had no idea of what the design constraints might be. But even as I was speculating, Henry Horn at Princeton was proving that there was indeed such a mechanism, and was showing how it worked.

If a tree has to grow up in shade so that there is hardly enough light to work its factories, it must obviously spread its leaves so that there is not one chink in its canopy to let a precious morsel of light through. The tree should be the shape of an umbrella. Horn was able to show that, for all their irregularity, this is indeed the general shape of trees of the climax forest. They are green umbrellas on long stalks.

But not all trees are umbrellas. Trees that grow in the open have branches bearing leaves all along their lengths at roughly regular intervals from the ground to the top. Horn explains this. If the top layer of leaves on such a tree were a complete umbrella they would intercept all the light, most of which would be wasted. The factories in the leaves would work flat-out, but this would not be very fast because the carbon dioxide supply is limited. I described this dilemma of plants in Chapter Four. But suppose the tree made an umbrella that was half holes! The umbrella would yield only half the amount of sugar but there would be light streaming through ready for the tree to make another umbrella half-full of holes a few feet lower down. The leaves of this umbrella have their own carbon dioxide supply and would still have light to spare. The sugar from the two half-umbrellas would equal what could be got from one whole umbrella, and there was still light pouring through the second set of holes to let the tree make a third half-umbrella another few feet lower down still. The sugar made by this layer is pure gain. And below that umbrella is another, and so on until the filtered light grows too dim to justify one layer more.

This is why trees have small leaves instead of large sheets, and why there are spaces between the leaves. It is why trees of the open places have branches at intervals up the trunk instead of just at the crown. Trees are installations for collecting and diffusing light so that as many as possible of their slow-running factories (leaves) can be operated together at the right light intensity and so separated that they have independent supplies of carbon dioxide gas. All this Horn has shown us with his brilliant musings and measurements in the New England woods. His book, *The Adaptive Geometry of Trees*, will surely be a classic of biology.

For the succession story, Horn's work clears up the final mystery. The climax trees are made more or less like an umbrella because there is not enough light in the shade of the forest floor to justify layers of leaves. The pure umbrella design is the most efficient for work in the shade and is absolutely necessary when a tree grows under its parents' canopy. But it would not be efficient in bright light. Yet the umbrella-shaped baby grows up to be an umbrella-shaped adult canopy-tree. For this job it is not efficient. Which is why trees of the climax forest can be less efficient than plants of all the communities which have preceded them in succession.

The extreme simplification of this account should be noted. All real forests are patchworks in which the patches are out of phase. Real trees die untimely deaths; real forests are cut by storms; real canopies are opened by disease and accident. Although the strategy of an umbrella-shaped monolayer apparently wins out for the dominant trees, there are plenty of opportunities for other plants to find a place in the patchwork of light and shade that is a real forest. This results in layering by different species of tree, by saplings and by bushes dwelling underneath the canopy trees. The total effect in an actual forest is likely to be a stack of layers composed of many species, each individual taking advantage of some imperfection in the umbrella of the canopy to snatch a morsel of light. Although a pure stand of climax trees would not be as productive as a pure stand of early succession trees built on the multilayer plan of several perforated umbrellas, the whole forest would probably produce as much, even in the climax stage. The most recent measurements on wild vegetation show that this is probably the truth of the matter. The climax community in forests is as productive as any other community, even though pure stands of climax trees are not. It is monoculture foresters who must worry about this problem, not ecologists.

We can now, therefore, explain all the intriguing, predictable events of plant successions in simple, matter of fact, Darwinian ways. Everything that happens in successions comes about because all the different species go about earning their livings as best they may, each in its own individual manner. What look like community properties are in fact the summed results of all these bits of private enterprise.

# Chapter Thirteen. Peaceful Coexistence

IF simple animals are kept in laboratory containers designed to meet their every need, they reproduce excellently. If three or four protozoans are placed in tubes of nutrient broth, a week or two later there will be thousands in each tube. A pair of fruit flies put into a milk bottle that has been well supplied with banana mash will soon turn into hundreds. So will a few waxmoths put into boxes of wax or flour beetles in dishes of flour. There is even a similar result if you start with a pair of mice in a six-foot enclosure which you keep supplied with endless quantities of food and water, though this exercise will lead you into social difficulties if you share your laboratory building with other scientists, there being no such thing as a mouse-proof enclosure.

All these demonstrations of natural fecundity run the same course. The animals reproduce very quickly at first, evidently making babies as fast as the mechanics of the business will allow. Within a very few generations the numbers are bounding away, faster and faster, so that the population grows like a cloud of exploding gas. But then something happens. The rate of increase begins to slacken and a daily census will show fewer and fewer youngsters added to the multitude. Eventually, the daily census will show little change. The numbers in the pressing crowd remain constant.

These epic histories of changing numbers are natural

grist for scientific man. He reaches for graph paper almost unconsciously and sets to plotting numbers against
time. The curve he traces is always shaped like the letter
"S," but drawn rather thin so that it does not double
back on itself, thus ∫.

When there are only one or two pairs of animals in the
cage, they rear magnificent families and the population
rises. To describe these early days we start tracing the
climbing curve at the bottom of the "∫." But soon there
will be not one or two breeders, but one or two dozen
of them; and they still raise magnificent families. Now,
the multiplied numbers are themselves multiplied, the
population starts to bound away; the graph climbs
rapidly up the straight section of the thin "∫." But finally
there is that apparent decline in the breeding effort
which damps the increase and turns the trace of the "∫"
toward the horizontal at the top. So scientific man says
that these population histories are "∫-shaped" or "sigmoid," which means the same thing in Greek.

It seems obvious that what we see in these ∫-shaped,
or *sigmoid*, histories has something to do with crowding.
In the early days plentiful food, lots of space, and hardly
any neighbors created good times for the breeders. We
expect large families, and we find them. In the next generation, and the next, times are still pretty good, and the
baby factories work at the same fast rate. But soon all this
breeding has raised a crowd. We would expect life to be
more difficult when this happens. Putative parents
would have to compete with others for food and spend
their energies in strife. There would be less calories for
reproduction, which should reduce the recruitment of
young into the population. And perhaps the harassed life
even of adults in that crowded place can be too much for
them, so that some die untimely deaths. Crowding
might thus both cut the birth rate and raise the death

rate at the same time. This seems the common-sense explanation of what is happening when the growth of an experimental population slows down.

So we put forward a general hypothesis to explain the sigmoid histories of laboratory populations. Populations grow until numbers crowd upon resources, after which the effects of crowding, notably competition for a limited food supply, press upon every individual. Then they breed less and die more, finally achieving an average level of misery that lets individuals barely subsist in a population that holds steady.

This general hypothesis is almost certainly correct for laboratory animals. The crowding required by the hypothesis is certainly real in the later stages of population growth because we made sure that it was. We forced the animals to struggle together for the right to live, which gives a certain plausibility to the hypothesis that they did so struggle.

Fruit flies, flour beetles, and mice are among the favorite laboratory animals, and whole lifetimes of work have been spent finding out what happens to them in crowded prisons. Female fruit flies, for instance, lay fewer eggs when they are hungry, and they get less to eat when they are in crowded bottles where the surface of the banana mash is crawling with flies. Experiments spanning some twenty years were needed to demonstrate this subtle point. There is also evidence that the maggots and pupae tend to die in mash wriggling with other maggots. Some more years of work were needed to show that point beyond the possibility of dispute.

Early on, crowded flour beetles were discovered to eat their own eggs by mistake as they blunder through the flour, making it mathematically certain that they would eat an ever larger proportion of their own eggs the more of them there were. The blundering beetles also bit into each other and into their own maggots, and their blind wanderings forced upon them a novel version of the ancient triangle problem; a pair seeking to copulate in a corner of the crowded flour would often not get the job done before being bumped by a third. People are still sieving flour beetles out of flour in many of the world's laboratories to make their daily counts, and they continue to discover fresh inconveniences suffered by crowded beetles.

Students of mice find even more spectacular results. In hopelessly crowded compounds, the complicated social life needed to sustain a mouse mother collapses so that she cannot look after her babies anymore. Crowded male mice can develop unsatisfactory symptoms too, ranging from orneriness to a state of sullen shock. The mice seem to be telling us something about human society, a message that students of beetles in flour find harder to get across.

The conclusions of these studies seem reasonable enough, and the lethal effects of overcrowding are both real and obvious. We can say that we truly know why population growth in the laboratory is always sigmoid and how the populations level off when the crowds become really dense. But it is not so obvious that this knowledge helps us to understand what is happening in the real world. The animals out there are not held in prisons. Their environment is not maintained at some arbitrary state of comfort. In addition, they must cope with the activities of other kinds who would share their food, and with hunters who would eat them.

Yet the mathematics of sigmoid growth has provided one of our deepest insights into the workings of wild nature, for all of the artificiality of the original model. It led directly to the formal description of that "struggle for existence" of which Darwin wrote and to the understanding of why there should be distinctly different species of plants and animals. And it even led to the most satisfying of all the lessons of ecology, that many animals and plants in fact live their lives largely free from the pressures of those deadly struggles.

These understandings came about because the crowding hypothesis lets one do some provocative algebra. The hypothesis shows clearly, in words, what the terms in a growth equation must be. The population will be boosted by the breeding efforts of the animals, which boost must be a function of the number $N$ of animals there, and by the rate at which a pair can reproduce in ideal circumstances; what ecologists call "the intrinsic rate of increase," or $r$. The population will be checked, according to our hypothesis, by how crowded it is. We say that the ultimate crowd represents the largest number that can be carried in the container, and we call this the "carrying capacity," $K$. In the slang of ecology the word "kay" is often heard when someone means some theoretical carrying capacity, just as the word "aar" means the breeding ability, $r$.

By balancing the breeding effort by a term that brings growth to zero when the number of animals equals $K$, a neat little equation results that describes precisely what is meant when talking about the growth of a population being brought under control by the effects of crowding. By rewriting the simple algebra in the calculus, one obtains an equation that can easily be solved for all possible numbers of animals, and that, for instance, could be fed into a computer that would draw a nice sigmoid trace on its XY plotter. More importantly, one can play games with the equation thus derived. $P$ can be a predator that kills with an efficiency $Q$, or $H$ can be a hurricane that slaughters a proportion $s$ every $y$ years, and so on. New more complicated equations might predict what the effects of such predators or hurricanes might be. Many of these mathematical games produce answers so unreal that we forget about them. But one such game led to the most remarkable of our truths.

Equations were set up to predict what would happen if two different kinds of animal had to compete for the same food in the same container. Intuition tells us that what ought to happen is that half the final crowd (more or less) is made up of each kind of animal. Each individual in that final crowd would have to struggle not only with its own kind but also with the other kind. If the two kinds were evenly matched and able to struggle strongly for their rights to life then, surely, the outcome of competition between two different kinds of animal should be a sharing of the food and equal numbers of the two kinds would result. The mathematics, however, do not predict this common-sense outcome at all. The mathematics predict total annihilation of one of the competitors and total victory for the other.

This prediction has been borne out in laboratory experiments. Competing animals will not coexist. The decisive experiments were done by the Russian biologist, G. F. Gause, at Moscow University in the grimmest days of Stalin's time. What he did was to set up numerous contests between different species of the small protozoan *Paramecium* to see if he could fault the predictions of the mathematics.

Gause kept his paramecia in the glass tubes of a centrifuge, which let him spin them in the machine each day to force the animals to the bottom while he poured off the exhausted food solution in which they lived without losing any animals. He could then top up the tubes with fresh nutrient broth. Any one of the common species of *Paramecium* would live alone very well in these tubes. From the eight individuals Gause put in at first, a thriving population of thousands would grow and this final number would remain constant for as long as he cared to spin them out daily in his centrifuge and replace their food supply. Eventually, Gause had several species at hand that he knew could live well in his tubes, that demonstrably thrived on the same food, and that were so similar that it was hard to tell them apart. If two of these species were placed together in the same tube and allowed to crowd, they must willy-nilly compete for that daily finite dose of nutrient broth. Thus we should see if common sense or the mathematics were right, if the species took the strain like Titans in unending struggle or if one scored total victory with the extinction of the other. The results were absolutely conclusive. There was total victory. The mathematics was right.

No matter how many times Gause tested two chosen kinds of *Paramecium* against each other the outcome was always the same, complete extermination of one species, and always the same species. Both populations would do well in the early days when there was plenty of room for all, but, as soon as they began to crowd, the numbers of the losing species would reverse their increase and go into a long decline that eventually left the winning population in sole possession of the tube. Gause could see this deadly struggle going on before his eyes day after day and always with the same outcome.

There are two obvious gut reactions to these results: one is amazement that what we expected to be a permanent struggling balance in fact became a pogrom, and the other is wonder at how the losing species can exist at all. This second thought holds the key to the whole affair and leads us to know that Darwinian struggles in the real world mean neither endless fighting nor deadly massacre but muted struggle.

The various kinds of paramecia live together in nature;

thus there must be circumstances in which the outcome of one of Gause's set-piece battles would be reversed. Gause knew enough about paramecia to guess at some of the ways in which this could be and was able to reverse the outcome of one struggle by a minor change of technique. Like other protozoa, paramecia were known to secrete chemicals into the water that were toxic to other animals; they were inclined to live by chemical warfare. But when Gause changed the water each day, he removed any such chemicals. So he tried leaving most of the water in and topping up with nutrient concentrate instead of changing the whole broth daily. In one of his series of experiments this was enough to reverse the outcome; the animal that had before always been the winner was now always the loser.

Then Gause stumbled across an even more revealing history, for when he tried yet another pair of species of *Paramecium* against each other neither became extinct but went on living indefinitely together in the tubes. When Gause looked at the tubes closely, he found that one species of *Paramecium* was living in the top halves while the other species lived in the bottoms. These kinds of *Paramecium* had found unconflicting ways of life possible in even those simple glass tubes of broth; they avoided competition by dividing the space between them. What obviously had happened was that one kind concentrated toward the bottom, where migrants from the other were overcome, whereas the second kind tended to swim toward the top, where their superior concentrations let them win. Stragglers into the wrong habitat space did indeed suffer deadly competition and struggle, but those of the majority who stayed where their own special strategy for life usually directed them were safe except from competition with their own kind.

There have now been many other experiments like

those of Gause, using many different kinds of animal and plant. They all result either in total victorious annihilation or in a sharing of the habitat in ways that prevent competition. Both the mathematics and the experiments, therefore, show that continued strong competition between different species is impossible. The different kinds must be kept separate. This at once leads to a splendid comprehension. Animals and plants in nature are not after all engaged in endless debilitating struggle, as a loose reading of Darwin might suggest. Nature is arranged so that competitive struggles are avoided. This is apparently why separate species are the result of natural selection. A species lives triumphant in its own special niche from which none can displace it. Only the stragglers into the niches of others must be removed by brutal struggle. Natural selection designs different kinds of animals and plants so that they *avoid* competition. A fit animal is not one that fights well, but one that avoids fighting all together.

Gause declared that his results illustrated a general principle: that no two species could live together indefinitely in the same niche, or, more simply, "one species: one niche." We call this the "exclusion principle," because the owner of a niche excludes all others from it.

When the exclusion principle was first heard of, it sunk swiftly and satisfyingly into the consciousness of all who studied real animals and plants. It was right, it reflected what scientists had felt all along. Every species was so distinct that we could even tell them apart by their shapes, by their stuffed skins and pressed leaves in museums. These shapes reflected their function. Talons mean a carnivore, hooves mean fleetness of foot, opposing thumbs mean climbing trees. Every museum man knew he was cataloguing functions, which is to say niches, when he catalogued the shapes into species. The deduction of niche from shape is what paleontologists do every time they reconstruct the life of an extinct animal. Animal and plant species were unique; they reflected unique niches; one species, one niche.

Field biologists found the new principle even more to their liking than did the museum people. It came to serve as a guide to their labors, a working rule that directed their patient watching as the work of a physicist is directed by his general principles of conservation of energy and mass. Whenever we find rather similar animals living together in the wild, we do not think of competition by tooth and claw, we ask ourselves, instead, how competition is avoided. When we find many animals apparently sharing a food supply, we do not talk of struggles for survival; we watch to see by what trick the animals manage to be peaceful in their coexistence.

Two species of cormorant are well-known in Britain: the common cormorant and the shag. These two birds look remarkably alike. They live on the same stretches of shore; they both feed by swimming underwater after fish; they both nest on cliffs overlooking the sea; they are both common; they were both hated by fishermen for stealing their livelihood. This last circumstance provided the eminent British birdman, David Lack, a chance to use the cormorants for one of the first tests of the exclusion principle in the wild. The fishermen had grown so vehement in their denunciations that local councils put a price on cormorants' heads and they were shot by the thousands. When it became apparent even to the fishermen that the slaughter made not a bit of difference to the fishing, the councils decided that their money might be better spent paying fisheries' biologists to report on the diet of the birds. The Plymouth marine laboratory undertook the investigation, examined stomach contents, and made field studies. Shags, by far the most abundant of the two species, ate mostly sand eels and sprats, which were not commercial fish. The common cormorants ate various things, particularly shrimps, and including a few small flatfish but no sand eels or sprats. The flatfish were commerical species, but the take was negligibly small. Thus the fishermen's wrath was quite unfounded. This was no surprise to ecologists, but the data were very much to the point for Lack's study. The food of the two species was obviously totally different. They were able to avoid competition, and the exclusion principle was upheld.

The fisheries' study showed further how the catching of different fish was ensured. The shags did their fishing in shallow estuaries, while the common cormorants went further out to sea. Their family affairs were kept separate too, for shags nested low among boulders or on narrow ledges, whereas the common cormorant nested on the high tops or on broad ledges. In short, these closely related birds, so similar to look at, had quite distinct niches. In their normal lives they were unlikely ever to come into competition.

Three yellow weaver birds of the genus *Ploceus* bred side by side in one colony stretching nearly 200 yards along the shore of Lake Mweru in Central Africa, and the man who found them promptly shot a few to see what they were eating. The stomachs of one species had hard black seeds in them, those of the second soft green seeds, whereas those of the third held nothing but insects.

The great variety of plants provides herbivorous animals with easy chances for food specialization. The effects of such food specialization are particularly obvious among herbivorous insects. Any amateur lepidopterist knows that caterpillars can be raised successfully only on the right food plant, which implies that the closely related butterflies that flutter together in clouds over a meadow are avoiding competition by doing their growing as larvae on different kinds of meadow plant. The plants force this specialization with specialized chemistry that makes them inedible to all but the specialist herbivore.

The engaging little birds called warblers are so similar, even in coloring, that learning to distinguish all except breeding males is one of the trials of a beginning bird watcher. In the eastern United States there appears each spring a particularly frustrating assortment of warblers flying north from their Caribbean winters, along common flightways, to their common breeding grounds in the woods of New England and eastern Canada. Five species, in particular, nest in the spruce forests of Maine and Vermont. The five birds are closely related, and the vegetation in which they breed is without obvious variety, just ranks of spruce trees. The beaks of the birds are all the same size, and alike, suggesting they can eat the same food. Investigations of enormous numbers of stomach contents by forestry people (who were looking for enemies of the spruce budworm) have shown that their food is, indeed, roughly the same. Although the food of closely related sympatric cormorants and weaver birds had been found to be different, these little warblers even had similar tastes for food. How then can they occupy different niches? How can there be more than one species of them? One of the foremost ecological theorists, Robert MacArthur, earned his doctorate by answering these questions.

MacArthur spent many long hours in the springs of several years watching the little birds. Each time he saw one, he noted exactly where it was: on top of a tree, at the side of a tree, on the ground, flying about. A stopwatch enabled him to measure in seconds how long the warbler stayed where it was. This was a tedious, time-consuming undertaking, for the little birds are hard to see in the dense spruce forest, and they never remain in sight for more than a few seconds. It took MacArthur two summers to amass a total of just 4 hours, 22 minutes, and 54 seconds of fully documented behavioral observations. But this four-and-a-half-hour's worth was enough for him to be certain of where each kind of warbler spent most of its time. And it was clear that the warblers worked in substantially different parts of the trees. One species spent nearly all of its time on the pointed spruce tops, another one lower down, a third near the ground, and so on. The spruce budworms, the most abundant food for all the warblers, lived all over the spruce trees, but the warblers hunted in their own special preserves.

Being such mobile creatures, the birds did poach on each others' space somewhat, but MacArthur was able to show that other behavioral traits stopped them from poaching many of each other's caterpillars even then. He timed the motions of each kind of warbler, noting how long each spent hovering, running along branches, or slowly plodding, and showed that each species had a characteristic pattern of doing things. One was more active than others; another was more deliberate. There seemed little doubt that these different activities reflected different hunting methods. One kind of warbler got caterpillars on tops of needles, another kind got caterpillars hidden under needles, and so on. Even though a warbler might poach upon another warbler's space and hunt the same kind of caterpillar there, the two might still not be competing because the hunting method of each ensured that each caught a different portion of the total crop.

It has long been recognized that the many grazing animals of the African game herds must be specializing in food, and students of the herds are now beginning to show how the niches of each species in those moving masses differ. All eat different bits of that immense pasture which is the savanna. Zebras take the long dry stems of grasses, an action for which their horsy incisor teeth are nicely suited. Wildebeest take the side-shoots of grasses, gathering with their tongues in the bovine way and tearing off the food against their single set of incisors. Thompson's gazelles graze where others have been before, picking out ground-hugging plants and other tidbits that the feeding methods of the others have both overlooked and left in view. Although these and other big game animals wander over the same patches of country, they clearly avoid competition by specializing in the kinds of food energy they take. And on the flanks of the herds move the specialized carnivores, the cats of different sizes, the hyenas on the watch for the suitably weak, the pack dogs that panic herds to cut out calves. All the exciting list of animals in the great herds can be seen by an ecologist in terms of the exclusion principle: a set of species which represents a set of niches, each one of which is a way of life conditioned to avoid competition with the other ways of life around it.

Peaceful coexistence, not struggle, is the rule in our Darwinian world. A perfectly fashioned individual of a Darwinian species is programmed for a specialized life to be spent for the most part safe from competition with neighbors of other kinds. Natural selection is harsh only to the deviant aggressor who seeks to poach on the niche of another. The peaceful coexistence between species, which results from evolution by natural selection, has to be understood as an important fact in the workings of the great ecosystems around us. It is also, surely, one of the most heartening of the lessons of biology.

# Chapter Fourteen. What Hunting Animals Do

PROBABLY the first man who ever kept sheep lost some to wolves, and cursed the wolves accordingly, passing on his opinions to his offspring so that wolves finally became ogres in fairy tales. In Europe one of the duties of feudal lords was to hunt wolves, and they did it so well that no wolves are left. In Alaska, agents of the government are still killing wolves from airplanes, with the declared belief that it must be "good" for the "game" to do this. No doubt part of this attitude stems from fear of the wolf, one of the few animals capable of attacking a man who does not hold a firearm; but our belief in the killing powers of predators extends even to those that cannot hurt people. We tell preschool children that cats are necessary to "control" mice; while spiders are "good" because they eat "flies." In western American states, comic cowboys have been shooting bald eagles from helicopters, giving that age-old excuse of the wolf-killer that the birds were taking their sheep.

Naturalists tend to frown on shooting wolves from airplanes or birds from helicopters but they find it hard to escape from the underlying philosophy. Predators kill prey, but if they kill it all, they will themselves starve. On the other hand, they will obviously kill as much as they can get. There must be a "balance" between their efforts to kill and the efforts of the prey to escape, a balance that controls the numbers of both predators and prey. We have learned that the supposed struggle between species is a decidedly muted affair that leads to peaceful coexistence; might not the struggle between predators and their prey be decisive in providing that general balance we see in nature? We are inclined to like this simple idea, but the truth is not so simple.

It is easy to think of a fierce hunting animal such as a tiger or a lion, or the even fiercer combination of wolves hunting together in a pack, as a fearsome scourge for their timid prey. The meek may escape by flight, but it is certain that the hunter must manage to fill its belly at regular intervals or the big cats and wolves would not exist. We can easily think of their hunting as depredations, as shepherds have always thought of the activities of wolves. And yet the few careful studies that we have of the last big cats and wolves tell a very different tale.

Adolph Murie long ago watched the wolves on Mount McKinley, living for years in the wilderness, recording what the wolves did, and giving us our first impartial study of what big predators really do. A very important part of the wolves' food supply was wild sheep, and Murie watched the wolves as they hunted. So here was the shepherd's curse truly at work on sheep that were without the protection of a shepherd. And Murie, in his painstaking way, deciphered a very special record of what the wolves killed.

In the Arctic, before finally decomposing, bones lie about on the frozen ground for years, particularly the hardest parts such as the tops of skulls, and on Mount McKinley there were many whitening skulls of sheep. Murie collected all he found, 608 of them. His years of careful watching told him that the only important cause of death for a Mount McKinley sheep (other than by a human hunter who would take the head) was being killed and eaten by wolves. So these 608 skulls represented a large sample of the wolves' victims. This in itself would not be very informative, but Murie was able to tell the age of each sheep when it was killed by the wolves from growth-rings on the horns. There were only two age classes in the collection of skulls, the very old and the very young. Apparently these were the only sheep the marauding wolf-packs caught, the infirm old and the feeble young.

Mount McKinley wolves did not kill sheep in the years of their prime; the skull collection showed this very clearly, and it was also quite consistent with Murie's personal observations of the wolf pack at work. If a pack of hungry wolves is the terrible instrument for destruction described by folklore and fable, this is not what common sense would expect. The wolves did not kill sheep in their prime, which leads us inexorably to the conclusion that they *could* not. Apparently natural selection has so fashioned sheep that they can outrun, outclimb, or outwit their formidable adversary.

More recently another pack of wolves has been watched at their hunting, this time on Isle Royale, an island forty miles long in Lake Superior. The only big game on the island that can feed the wolves through the winter are moose, and it is the powerful moose that the sixteen or so wolves of the Isle Royale pack hunt down through the winter snows. The pack has an appetite that requires one moose every week. From the air David Mech watched to see how they got this moose, picking up the clear trail the wolves left through the deep snow in the early morning, then winging overhead as they went about their hunting. Sixty-nine times he followed them thus. Nine times he was up with the hunt all the way from the find to the check from the check to the view to the kill. Twice he saw the kill near where he could land his plane, came running at the pack waving his arms to drive the hungry wolves from their meal, and had a look at the carcass himself. Studying chewed-over remains of many more kills, he saw very clearly what happened when the wolf pack closed with a moose.

In the chase itself the wolves were superb. When they had once hit on the trail of a moose there was very little chance of the moose avoiding them as a fox so often avoids fox-hounds. Perhaps this is not surprising, for the trail of a moose running in thick snow would not need much trailcraft in the following, but most of the moose escaped with their lives all the same. The wolf pack either gave up early or after a short skirmish with a moose that had turned and stood to confront its persecutors. Whenever a prime moose chose its place and stood to fight, the wolves gave up and went away. The moose they closed with and dragged down were always youngsters in their first two years of life, old senile moose, or the sick.

Both Mech's observations of the hunting and his examination of the remains showed very clearly that the Isle Royale wolves never took prime moose. Like the Mount McKinley wolves, they took only the easy meat from the herd: the old, the young, and the sick. It is easy to see why the wolves leave prime moose alone; they are too dangerous. There can be little doubt that if sixteen wolves really closed with any moose, they would overcome it, however strong and fit it might be. But some of the wolves would get hurt, and a hurt wolf can hunt no more. Natural selection sees to it that the strain of brave aggressiveness in wolves is purged from the wolf gene pool because such individuals would incur more than an average share of being fatally hurt and thus would leave fewer descendants. The wolves that have survived the winnowing of natural selection are those that make do with the prey they can kill without danger to themselves.

Since packs of wolves habitually kill neither prime moose nor prime sheep (obviously for different reasons) our preconception that they might regulate the numbers of their prey is bound to develop doubts. The wolves certainly have some effect on the populations of their victims because they kill some of the young, but this is much less than the depredations that folklore and intuition would lead us to expect. And for other big predators, which hunt alone, the difficulties of severely culling the numbers of their prey are even greater.

The American mountain lion, sometimes called a puma, a cougar, or even a catamount, is small as big cats go, but it is still a powerful animal and it is known to hunt white-tailed and mule deer. The accounts of mountain lions in popular mythology might make it seem as terrible a scourge to the "defenseless" deer as wolves were supposed to be to sheep, yet the reality is again very different. We still have no good eyewitness accounts of much mountain-lion hunting, partly because the lions are secretive but also because our philosophy of killing them has made them nearly extinct over most of their old haunts. But there are some left in Idaho, and M.G. Hornocker recently won a Ph.D. with some remarkable tracking and woodlore in the haunts of the mountain lions.

Hornocker found that the Idaho mountain lions in winter are complete loners; each has a tract of wilderness through which it hunts alone. Tracks in the snow showed that this loneliness is from choice, because a lion will turn away from the tracks or the presence of another. Even powerful males in their prime will turn aside from a weaker or younger animal. There is no social dominance in this, no expulsion of a weak animal from a superior's preserves. Hornocker concluded that the lions' habit of each keeping to itself had been preserved by natural selection because of the difficult task of hunting. The big cats could only kill deer if the deer were quite unsuspecting. Deer that were nervous because a marauding lion had been through the country were virtually unattackable by another. Although there were plenty of deer in the wilderness, the mountain lions had to keep a very low profile or they could not catch deer. This does not sound as if mountain lions kill easily or that they have much influence on the numbers of deer in a wild population.

The really big cats are less than impersonal killing machines too. George Schaller relates the killing methods of tigers as they take tethered domestic buffaloes and his account does not suggest that killing is safe or easy for a tiger even with these advantages. The tigers ran at their prey, half-climbed on their backs, wrestled them to the ground, then dodged the flailing hooves to seize the buffalo by the neck. It always took several minutes for the buffalo to die. This was not at all the quick surgical operation of killing that nature-story accounts of the big cats would lead us to expect. If it has that much trouble with a tethered domestic buffalo, it is not improbable that a lurking tiger might normally let the formidable animals pass unmolested and look for something more out of sorts.

It is probably generally true that large vertebrate predators go about their killing cautiously. Whether it is a lion or tiger stalking a game herd on the plains, or wolves running down their quarry through a northern winter, the predator always faces the reality that it must kill again and again if it is to survive. Fifty-two desperate encounters a year would be likely to result in hereditary oblivion. Neither big cats nor pack-hunting canines have the firepower to pull off fifty-two safe butcherings a year if they attack the fit and the strong. They avoid desperate encounters, unless extreme hunger drives them. Usually they feed by culling the old, the sick, and the young.

There is no doubt that all these big fierce predators have some effect on the numbers of their prey because they kill the young. But they cannot usually kill a very large proportion of the young because the number of predators is relatively small. The young typically make their appearance at only one time of the year, and the predators must live the rest of it too. The numbers of big cats and wolves that a herbivore mother must look out for in the spring is mercifully low because it will be the number that has been kept alive through the winter by the supply of old and sick animals.

It thus seems very likely that the larger and fiercer predators are not nearly so important in regulating the numbers of animals in nature as common sense suggests. They are really to be looked upon as scavengers without the patience to wait for their meat to die. They cheat the bacteria who would have got the bodies otherwise. Two rather pleasing thoughts come from this discovery. One is that the lives of big game animals are lived in a large measure of freedom from the awful world of tooth and claw that we can conjure up by a careless reading of Darwin. Not only do these animals live in that peaceful coexistence with their neighbors, which the mathematical ecologists discovered, but they also may live with less fear of being killed than we had supposed, except as a sort of euthanasia. The second pleasing thought is that those who like to shoot big game themselves no longer have a pretext for killing off the wolves and cats before they start on the deer.

But if the firepower of a big cat is insufficient to devas-

tate a herd of game, the firepower of the smaller predators may be truly awful. A spider or a wasp is a deadly
efficient engine of destruction. Perhaps most of the
species of hymenopteran insect that we loosely call
wasps are in the business of hunting caterpillars and
grubs of other insects, piercing them and laying their
eggs under the skin, letting the maggots feed and grow
on the living flesh of their victims, and eventually flying
away from the empty carcass as mature wasps themselves. Although the victim thus takes long to die, the
crucial predatory act is the initial attack by the female
wasp on the caterpillar, and in this encounter the caterpillar stands no chance. When a wasp strikes, it is not
like a tiger striking a buffalo; the issue is never in doubt;
the chance of the wasp's being wounded is zero. The
same must be true when a web-spider closes with a fly
struggling in its meshes. It must also be true when a
spider-hunting hornet plunges like a dive-bomber, with
its armor-plated body and its poison-loaded stinger, on a
spider sighted in the open. It must also be true when a
tiger-beetle pounces, when a praying mantis reaches out
with its dreadful arms, and when a large carnivorous div-
ing beetle finds a small tadpole. In all these, the only
hopes for the hunted are to escape detection or timely
flight. We might expect, therefore, that small predators
can have more potent effects on their prey than do large
predators.

That these small predators can truly be devastating
has been shown by the success stories of entomologists
when they have ridded farmlands of an agricultural pest
by introducing a suitable natural enemy, so-called
biological control. Celebrated among these successes is
that of the Californians who ridded the orange groves of
the little, white, flightless insect called the cottony-
cushion scale, which had appeared in plague-like pro-

portions in the 1880s, threatening to destroy the entire citrus industry. The cottony-cushion scale was an Australian insect that must have come to California by sea in a cargo of fruit, so a Californian entomologist went to Australia to look for enemies of the scale. He had wasps in mind and he found some, but they turned out to be ineffective. Then he found an Australian ladybird beetle called the vedalia, a little red ladybird with black spots like those common in Europe and North America. He sent to California a total of 129 live vedalias.

In January the few vedalias were put on an orange tree heavily infested with the cottony-cushion scales and the tree was covered with a muslin tent. By April the tented orange tree was free from scales but rich in ladybirds, and they opened the tent to let the beetles out. By July the whole orchard of 75 trees was free from the pest. The news spread, and planters journeyed far to collect the precious beetles for their own estates. Within a year the whole of southern California was rid of the plague of cottony scales.

This pretty ladybird, the vedalia, has proved itself to be a far more deadly predator than any wolf or tiger. It searches with diligence and kills with utter certainty. It processes the calories from the bodies of its victims into its own babies with such dispatch that the next generation is ready to carry on the killing in just twenty-six days. As we have seen, this ferocious attack can exterminate the prey in an entire country within a season. But what can the ladybirds do then?

Only part of the success of the vedalias was due to their deadliness and mobility; the rest came about because they were given a concentrated target. In the wild Australian home of both the vedalia and the scales, there were no citrus orchards, and the food base of both must have been scattered trees in the forest. Life in a colony of cottony-cushion scales on an isolated forest tree might well go on for generations before a flying beetle found the colony to begin its killing. And life for vedalia beetles who must hunt them would involve sending out the next generation in pioneering searches for new and distant trees bearing colonies of their food. The scales escaped their enemies by living scattered across the land, and their ladybird hunters got their livelihood by arduous and unremitting searches.

After the first slaughter in California it seems that something like the ancestral Australian pattern was established between the vedalia beetles and their prey. In later years the infestation was gone, but if you looked hard enough you could find a colony of scales somewhere in the orchards, but they were so few they were no longer a nuisance. Chance had let a few escape the attack of the vedalias and they served to found new colonies after the scourge of beetles had passed them by. Each colony would live until a wandering beetle found it, when it would be rapidly wiped out. But meanwhile another colony started up somewhere else. Life for both vedalias and scales became a game of hide and seek across the spaces of California.

It is likely that games of hide and seek between predators and prey run on indefinitely for many small species of animals. The outcome is a consequence of devastating killing power, and it can be expected on common sense grounds as well as predicted by formal mathematics. Scientists make equations which show the numbers of prey growing in the classic geometric way but which are cut back by the attacks of predators. In this formal scheme each attack results in a kill, as it will for small animals, and each predator turns its victims into more predators after a suitable time lag for the business of reproduction. The result is a model that predicts the absolute wiping out of the prey as the predator numbers build up, perhaps after some oscillations. This is what we see in nature. Locally the prey is wiped out as the model of efficient hunting says it must be, but the game has been started all over again somewhere else by refugees from the first game. The result is a scattered population of prey animals living many generations in security, but occasionally faced with local annihilation.

This pattern results even when the game is started on so uniform a board as was provided by the ranks of citrus trees in the equable Californian climate, but in nature there are many other forces at work to frustrate the hunters. The plant food of the prey is itself scattered, there are various physical barriers to both search and escape, and the fluctuating seasons, to say nothing of vagaries of the weather, influence the outcome.

In places of seasonal climate both predator and prey have to endure a hostile time, perhaps a winter through which they must persist in some quiescent state, as seeds, eggs, or dormant adults. It often happens that only small numbers survive this lean time. With every growing season, therefore, a new game starts, and this game has some of the qualities of a race. The few prey animals that have got through the winter set about the business of reproduction, probably helped by the lush spring growth of their food plants. But the predators will find little to eat and will not be able to produce many young until later in the summer when the population of its prey will have built up. The predators may not have time to build up devastating populations before the coming of the next winter clears the game board once more for a fresh start.

The lives of small predators and their prey are thus different in fundamental ways from the lives of large animals. Large predators live alongside and within sight of their prey, like prides of lions lying in the sun as the herds of African game wander by them. This is essentially because the weapons of the big hunters are not good enough for the safe pursuit of indiscriminate slaughter. But this peaceful coexistence is not possible for small predators and their prey, so they must live scattered, the one fleeing and hiding, the other searching and destroying. Moreover the large animals live through many different seasons, which lets them smooth out the effects of weather. Short-lived insects and their kind pass through several generations a year so that they meet the different seasons with different generations. The numbers of predators and prey can be differently hit by such adversities as winter. This means that the effective power of the predators is often nullified by the further scattering or reduction of populations. Large predators and prey persist in a harmony which owes much to a certain lack of weaponry. Small predators and prey coexist, if not in harmony then in relative safety, because the very deadliness of the weapons combines with chance and their short lives to keep the antagonists scattered and apart.

# Chapter Fifteen. The Social Imperatives of Space

WE probably know more about birds than about any other animals. Birds are attractive and easy to see. Laymen and naturalists alike know about their migrations, their courtship, their returns to ancestral breeding grounds in the spring, and the chorus of their songs in the early morning. Many of our feelings about nature are fashioned from these familiar doings of the birds. Certainly this knowledge must contribute to that comfortable feeling we have that there is a balance in nature.

Every year there are about the same numbers of birds about, and this is strange because we see them raise their families. At the end of the breeding season there must be many more birds than at the beginning, probably at least twice as many if we add up all the babies and their parents. But next spring only the same familiar number of nests are occupied. This has long been known to countrymen in a general way, but now we have a number of good censuses of breeding birds to show us that it is really true. There are particularly good census data from herons in England, storks in Germany, and titmice in Holland over spans of up to forty years. Every year the numbers breeding over wide areas are essentially constant, except for occasional years of depression, of which the cause is obvious. One such depression was of herons in England after the 1947 winter, one of the coldest ever in which dead herons were found beside frozen water. Vacant nests in the ancestral heronries in the following spring can easily be understood. For the most part, however, the census data confirm the beliefs of countrymen that roughly the same numbers of birds breed every year. This has not been easy to explain.

The problem may be simply stated. In spring and summer the numbers of birds grow, though we do not know by how much. But they certainly grow, and by different amounts each year depending on whether the breeding season has been good or bad. But when the next spring comes, essentially the original number of breeding birds turns up to breed. Something has apparently happened to the surplus produced by the previous year's breeding effort. This can be understood because winter has come between. The difficulty is that whatever happens in winter lets the same number through every year, despite the fact that each winter starts off with a different number of birds.

Our first thought is that a crowding process occurs in winter, similar to that in a glass tube gray with *Paramecium*, like those of Gause, or in flour crowded with flour beetles. Undoubtedly pressures of this sort are present in winter, when food is short. There might well be increased competition for food in winter, and refuges from the winter's violence might be harder to find when too many birds are looking for them. To this extent, winter survival becomes a function of the number of refuges available, perhaps the number of bramble patches, the area of bottom land out of the wind, and so on. But it is hard to believe that the effects of being crowded into these places would cull with such exquisite precision that they let the same number through each winter, year after year. Quite apart from the subtlety of the required mechanism, winters are not all alike.

Whatever pressure of crowding might work in winter, it could only work through killing. The pressures of crowding that worked to control flour beetles, fruit flies, or paramecia in laboratory glassware worked by curbing the birth rate as well as by increasing the rate of death, but winter control of birds can only work by killing because the birds are not breeding them. For this reason, some ecologists have put their faith in predators to do the culling. They have never been able to find field evidence that avian predators can work in this way, nor does a close look at the workings of predation in the wild, as in the last chapter, encourage the idea. So the problem remains. The number of birds turning up to breed at the spring census is constant from year to year, even though the number fluctuates widely from autumn to autumn. Although we can perhaps accept some tendency for the surplus to be culled in winter, it is extremely doubtful that winter's hostilities can account for all the regulation we see.

If we cannot find the answer in the winter, it seems logical to look for it in the spring. Suppose that the number of breeding birds was fixed in some way, by an environmental parameter perhaps; would this not account for our observations? Indeed it might. Our annual census is of breeding pairs. It is the number of breeding pairs that remains constant from year to year. With most kinds of bird we have no easy way of telling if there are any at all which are alive in the spring but which do not breed, let alone making a census of how many there are.

We do know of some species of birds where there are populations of juveniles that are not ready to breed, the long-lived sea-birds such as herring gulls being obvious examples. It is possible that these species delay breeding because long practice is needed at searching shorelines for gull-food before an individual can be good enough at it to provide for growing babies. But this kind of non-breeding population does not help us to resolve the problem of the constant numbers, particularly in the short-lived song birds, where there is very little reason to expect it. What the hypothesis needs is a population of birds capable and ready to breed but denied the chance. This is, on the face of it, an unlikely thing. Natural selection ought to be against it.

The only birds that exist do so because they are of lines that are superlatively good at making babies. Every individual must carry the genes for this. They must all get an equal chance to breed, and professional non-breeders will be ruthlessly suppressed by natural selection. So the idea of a set number of breeders that cannot be exceeded even if more birds are alive is something that many an ecologist has found very hard to accept. And yet we know of many habits of birds that seem to demand an allotment of space to each bird, to each pair, or to each group. We have come to talk of many of these habits as "territorial behavior." If for any reason at all the life of a bird, particularly a breeding bird, requires a minimum space or portion of real estate, then an explanation of the constancy of numbers in the spring is at hand; the numbers of breeding pairs would be a function of the amount of real estate available, and this is the same every year.

The idea that birds might "own" "territory" is an old one. Gilbert White of Selborn had something to say about it, but Olina writing in Rome in 1622 talked of the nightingale's "freehold" a hundred years earlier still. It would be surprising if forgotten naturalists of the Greek and Roman Empires had not remarked the same thing, to say nothing of our countrymen-ancestors back in the stone age. But the modern naturalist owes the basis of what we now know to the work of one man, Eliot Howard. Howard did some inspired watching of English birds fifty years ago and published his findings in what was to become one of the seminal works of science, *Territory in Bird Life*. As with other seminal ideas, the idea of the territoriality of birds has been used to support some queer propositions. It is best to forget what one may have heard of these things and to follow Eliot Howard at his birdwatching, to see what he saw, and to learn what territoriality really implies. He watched many birds and followed his thoughts through the seasons with bird after bird. We can learn what Howard found by following his observation of just one of them, a little English finch with a yellow head known colloquially as the yellowhammer.

The yellowhammer is a year-round resident of England. It is well-known and conspicuous, the male particularly, with his yellow head and his habit of sitting on fence-posts and gates by the country roads to sing a curious little song that goes "a-little-bit-of-bread-and-NO-cheese." In winter, yellowhammers live in flocks: males, females, and other kinds of birds as well. But as spring approached, Howard noticed that the cock yellowhammers began to edge out of the flock a bit, to spend a while sitting on a post, and to experiment with snatches of song. Soon the cocks would spend longer away from the flocks on their posts, and the cry "a-little-bit-of-bread-and-NO-cheese" became a common country sound. The cocks seemed impelled to their chosen perches by some obscure urge. Several times Howard, watching the flock through his binoculars, saw a yellowhammer suddenly leave and come plummeting back to his post, almost as if he had forgotten something. Once there he would sing "a-little-bit-of-bread-and-NO-cheese."

Eventually the flocks broke up entirely as all the birds

went their own ways. By then, all the cock yellowhammers had their favorite perches, did all their foraging near these perches, and would frequently pop back up to sing. Where once the males were parts of gregarious flocks, each was now by itself, each spaced apart from the others, each advertising its presence with song. And these separated males became irascible, rushing angrily at any other male that came near the favored spot. There sometimes ensued one of those fights in the air which seem so characteristic of birds in the spring. The display was always soon ended, one bird would fly away, and the other would return to his post and sing about bread and cheese.

At this stage in the proceedings the cocks acted as loners; aloof, solitary, irascible, showing off, and apparently eager for ritual combat with any other male, or indeed with almost any other kind of bird, which approached them. People who like to read human motives into the doings of animals have long marveled at these challenges, displays, and aggressive combats of the cock birds in the spring, and passed their own judgments on the event: these were the struggles of cavaliers for mates. The males sounded their challenges like knights of old before the downcast eyes of maidens and rushed to battle for the maiden's hand. Here was survival of the fittest at work, the physically fittest. Only the most perfect warrior should breed. This deformed view of what Darwin wrote has been the excuse of the worst military bullies of this century. But Howard noticed that these fights began well before any females were present. This was particularly noticeable in some other kinds of birds that wintered outside England and whose males arrived back some weeks in advance of the females. These cocks did their "fighting" when their hens-to-be were in another country.

The purpose of this singing and the belligerent attitude of the cocks in the weeks before the hens arrived must surely be judged by the result. Nobody got hurt in most of the fights, which were essentially ritual. Moreover the cock that had first taken up its position on the post invariably "won," in that it was the intruder who flew away. So aggression itself was not a purpose of the behavior. What really resulted from this spectacular play in the spring was that the cocks became nicely spaced out across the land, with each getting fixed into its brain that the neighborhood of a certain post was where it should be. This spacing and learning of home is the prime result of the behavior and so must be its evolutionary purpose.

But that the songs should, among other things, also attract females seemed reasonable enough, particularly for those migratory birds who were scattered over whole countries and whose late-arriving hens had to find the cocks. Sure enough, each yellowhammer was joined by a hen in the fullness of time. But the song must serve other functions as well, because the males went on singing even after they had acquired their mates. It looked as if the song might have lasting value to the pair.

Each pair built a nest and reared its young near the vantage point from which the male had made his first experiments with song. They foraged nearby in that same local space that he had come to treat as home. And they both became irascible and intolerant of other birds. Now both would fly at an intruding male or at an intruding female; and sometimes the matronly hen would fly at an intruding hen, attacking her in earnest.

To Howard, the advantage of this package of habits was clear. The singing and isolation of the males in the early spring resulted, when they were joined by their mates, in isolated, separate pairs. Each hen became used to the same surroundings that had evidently become fixed in her cock's senses as home, and the pair were kept together by this mutual sense of home. One obvious outcome of this was that the pair were bonded together; a form of marriage contract had been arranged. There are, of course, other mechanisms that birds can use to maintain the pair-bonds and many, like penguins and swans, seem to establish lifelong attachments. Yet any mechanism that served to keep together a flighty couple of migratory birds while they raised their young would obviously be of advantage to the breeding effort. So Howard suggested that one of the reasons that territorial behavior had been preserved by natural selection was that it cemented the pair-bond. The behavior both brought couples together and, through the intolerance of both cock and hen to third parties, maintained the breeding unit against the difficulties of eternal triangles.

But Howard stressed another implication of the behavior, which has come to seem more significant to modern ecological eyes: each pair of birds acquired a private foraging ground near the nest. This let them collect food for their young without wasteful journeys. Perhaps more important still, it assured them of the food in the surrounding spaces for their own use without losing any of it to competing individuals of their own kind. This must be of great benefit to the workings of a large-young gambit. In the early stages of territorial display physiological changes are going on in the birds that determine how many eggs will be laid, and it is at this same time that the perimeter of airspace in which another bird will be challenged is worked out in the course of repeated encounters. The establishment of a territory thus can well involve a ritual measuring-off of the food supply into something like family-sized allotments. This effect must certainly result in an increased chance for survival of the babies attempted. The genetic basis for the behavior would thus be strongly chosen by natural selection on this ground also.

It is possible to think of other advantages of the behavior: keeping birds on familiar ground where they can dodge predators; minimizing disease by minimizing contacts; giving private use of shelters; and so on. Howard discussed them all, and they are still argued about as we collect more data. Probably the provision of a private food supply is the most pressing of the benefits conferred, though the marriage contract may also be of signal importance. The important thing is that Howard showed the way to an hypothesis that completely explained the doings of the birds in the spring in ways acceptable to Darwinian logic. The singing, the displays, and the celebrated combats were all parts of a pattern of behavior whose advantage was to let the birds turn resources into babies with a high degree of success. The behavior conferred fitness. Any serious deviation from it could be likely to lead to poor performance at breeding and thus to hereditary oblivion.

Another consequence of the behavior has to be stressed here, however, because it has given rise to much misunderstanding and even to downright mischief. It is the circumstance that led Howard to call the behavior "territorial." This notion of "territory" has trickled out of the pool of jargon of the ecology trade into a broader English usage, where it has led to some dangerous wrong-thinking. The facts of the matter were that birds such as the yellowhammers came to hold space in which they fed, and that they dominated an area even to the extent of defending its boundaries against other birds of their own kind. But the proper way to look at these things is as side-effects of the behavior; inevitable, but still side-effects. The behavior itself was a pattern of sing, dance, and pursue, a set of signals that together said, "Keep away from me." The evolutionary advantage of the behavior was that food was reserved for a growing family even as the parents were held together. But the effect to the human eye was that there was a space around each pair of birds that looked remarkably like their property.

Given the limitations of language, it was an almost inescapable use of English to call these spaces in which the birds reared their young their "territories." Howard stressed as strongly as words would allow that the important thing was that birds sang and "fought" because this secured food for the young and union for parents, and this behavior could, for want of a better word, conveniently be called "territorial." Lesser people have seized on this catchword "territory" and said that birds sang and fought to claim real estate. But we have no evidence that they do. Territorial behavior is explained as a process of "keep away from me" not of "this dirt is mine."

"Keep away from me" is a two-dimensional demand. It requires spacing, and space is something for which there is an absolute limit on the surface of the earth. This at once prompts the idea that territorial behavior sets a limit to the numbers of birds that breed every spring. As the ethologist P. H. Klopfer puts it, "If the size of the territory cannot be reduced beyond a certain point, and if successful reproduction requires that the bird possess a territory, the regulatory function of territories becomes a function that is beyond dispute" (*Habitats and Territories*, Basic Books, 1969). To some ecologists this has been a more comfortable explanation of the constancy of numbers of breeding birds in the spring than that which relies on an exquisite winnowing of surplus youth by death in winter. But it is necessary to scrutinize the idea suspiciously before we close with its tempting comfort because there have been grave difficulties for it, both practical and intellectual.

The actual size of a bird's territory is plastic. Many naturalists have mapped the territories of song birds in the spring, and there is no doubt that the size of territories can be modified to suit local circumstances, including the numbers of birds wanting territories. In some other species of birds there is evidence of quite spectacular changes in the sizes of territories from year to year, which are clearly correlated with the food supply. Pommerine jaegers or skuas, for instance, big gull-like birds that nest in the arctic tundra and feed their babies on lemmings and voles, will build nests every few hundred feet in a year when the tundra is running over with the animals of a lemming "high," will have territories of many acres each when lemmings are in short supply, and will not bother to nest at all in the years when no lemmings are around. It is clear, therefore, that territory is not an inflexible arbiter of number. But still, territory could set an upper limit if there is a minimum size beyond which territories could not be compressed, and this is all that we need to add to the effects of winter to keep the numbers at spring census roughly constant in many places.

A second reason to be suspicious of the idea that territory determines the number of breeding birds is more serious. The hypothesis requires that there be a population of individuals that are denied a chance to breed. Individuals without territories will have no babies and so cannot pass their genes on to posterity. How, therefore, can natural selection allow such behavior to persist? This is the objection to all ideas of surplus non-breeding populations that I raised earlier in this chapter, and it is a very powerful argument. We can, however, get round it for this particular instance of a surplus population being unable to breed because they cannot win a territory. We assume that all individuals start with an equal chance to gain a territory, except that older, more experienced birds may have an advantage in having learned how to do it (and birds have very considerable powers of learning for all their lack of intelligence). Which individuals get territories is then pure luck, which is not unreasonable. And we further assume that no surplus bird could spoil the system by breeding without a territory because the behavior is absolutely necessary to the breeding process. This also is a reasonable assumption.

It is helpful to think of the outcome of a territorial encounter from the point of view of the loser. Like the winner he carries genes that say, "Go out and display, find a place you like and see that others keep away from you." But he also, again like the winner, has a program written by his genes that says, "Quit without too much fuss if your opponent comes on with that confidence that shows that he has been successful before." If the loser did not have this second program telling him when he ought to quit, what then? Perhaps he would fight for real, or at least persist in his assaults. But then his chances of breeding would be very poor indeed, probably zero. He might cause so much trouble that the breeding effort of the territory owner was ruined, but this would not do him any good. In surroundings of constant strife, he could leave no babies himself and his aggressive "don't quit" genes would be lost from the population. But if he does quit when faced with an encounter he obviously is not going to win, he yet has a chance to breed. He may find a territory elsewhere, or, and this is the likely thing, he may have learned so much from his encounters that he makes sure of a territory and thriving family next year. It is, therefore, to the advantage of the loser in a territorial bout to quit in time; quitting can be a "fit" thing to do.

Thus we have a plausible hypothesis to explain that portion of the balance of nature that is revealed by the constancy of numbers of breeding birds in the spring. After the effects of crowding during winter have brought the numbers to some sense of order, a final fine-tuning is provided by the fact that the physical size of breeding territories cannot be compressed beyond a certain point. This hypothesis implies that there must be years in which there is a considerable population of adult birds that are physically capable of breeding but do not do so because they have been denied territories. The test of the hypothesis, therefore, must be to try to demonstrate the existence of the required non-breeding population. The best evidence comes from a study which may well be called the work of the "Maine Gunners."

The scientists I call the Maine Gunners did not set out to investigate population regulation in birds and at first could have had no idea that they would earn fame with evidence that was to be crucial to an understanding of breeding strategies. They were employed to study a caterpillar pest, the spruce budworm, which was ravaging the coniferous forests of Maine. These spruce budworms were used as food by several kinds of warbler that were nesting in the spruce trees and raising their broods almost entirely on the budworms. I have already described how Robert MacArthur worked out the way that five species of warbler shared this copious food source. It seemed likely that these breeding birds might be having a useful restraining influence on the pest, and the Maine Gunners determined to discover this. Their plan was simple. They would choose two typical stretches of forest, remove all the birds from one stretch, and compare the fortunes of the pest caterpillars in the forest without birds with those of the forest where the birds were left alone. But before removing any birds they would, of course, count how many were there. In doing so they laid the foundation for their exciting discovery.

With a little experience, and with the necessary patience and pertinacity, it is reasonably easy to make an accurate count of the number of song birds breeding in a wood. Teams map the positions of singing males, follow them to their nests, map the nests, and so on. The Maine Gunners had chosen forty acres of forest for their experiment. They watched and counted for two weeks, and found that there were one hundred and forty-eight pairs of the warblers breeding in those forty acres. Then they came back with their shotguns and started shooting, their declared purpose being to kill every bird in the wood. They knew where each of the hundred and forty-eight pairs lived and expected the slaughter to be quick and complete. But it took longer than they had reason to expect. After three weeks they had killed three hundred and two cocks and a lesser number of hens. And there were still birds singing all over the wood. A hundred and forty-eight breeding males to start with; three hundred and two of them killed; and still some remained.

The gunners had killed more cock birds of breeding habit in the wood than were there to start with, and this must mean that extra birds were coming into the wood as fast as the original occupants were cleared away. This was the time of the full breeding season. If the extra birds had been breeding elsewhere, they would not be available as replacements for those that were shot. Apparently there was, indeed, a surplus population of non-breeding birds that had been denied territories.

The Maine Gunners well knew the significance of

what they had discovered. Some of them went back the next year to confirm their results, and we can imagine with what anxious care they made their preliminary count. This time they made the total one hundred and fifty-four breeding pairs. Then they took their guns again and shot three hundred and fifty-two males and a goodly number of females as well. Birds were still arriving in the wood when they quit. The evidence that surplus birds were available was thus overwhelming.

There was one oddity about the Maine results. Both times the gunners shot many more males than females, almost as if it was only males that were surplus to the breeding effort. But we have no reason to believe that the numbers of the sexes in these birds are unequal. It seems likely that the season had advanced too far for females to be any longer available. The breeding strategy of these birds made the females sexually active in the appropriate days of spring and by the time the gunners had let the woods fill up with breeding pairs, made their counts, and spent their days of shooting, the females were past their receptive time. The mating season for wallflowers was over, however much the males were still prepared for a chance at sexual display.

We have a few similar data from mass shootings, but not many because the work is not popular. A number of other, less violent studies, however, indicate that same surplus of birds to territorial spaces. One man spent many years following the fortunes of flocks of magpies in Australia and showed convincingly that there were always two subpopulations of magpies, a small tribe of birds that did the breeding and a larger, drifting flock of non-breeding magpies, whose members never found a breeding place until one was vacated by the death of a member of a tribe. We also have observations of various titmice and grouse in Europe that show that, for these birds, there is indeed a physical minimum size for a territory. These species cannot successfully pair and raise young unless they have a certain physical space. These various lines of evidence are probably sufficient to confirm the plausibility of the general hypothesis.

It is now possible to believe that the constancy of numbers of breeding birds from year to year is partly a consequence of a pattern of home-making that requires allotting space to each breeding pair. But the control of populations that results is an accident. This must be emphasized. Birds are not provided with a mechanism for regulating population; they are provided with a territorial breeding system fashioned by natural selection to boost the numbers of babies they can rear, not to restrict them. That this mechanism should also set a ceiling to numbers is a bizarre side effect.

Many of the more complicated patterns of behavior are likely to have side effects in a like manner. An obvious example is the social hierarchy that appears in many different kinds of vertebrate, such as the celebrated peck-order of chickens. A social peck-order must have survival value for the behavior would not have evolved otherwise. We can see some of the ways in which it confers its benefits. Chickens which have settled down in their social hierarchy are peaceful chickens that go about the business of collecting food without wasting their energies on constant disputes. When social order is enforced, all of them raise more babies, both the high status chickens and the low. A chicken at the bottom of the peck-order is yet better off than it would be under anarchy. But there is no doubt that the bottom of the peck-order in times of crowding can be a difficult place. If crowds grow dense the chicken at the bottom may be ejected from the society completely. This would probably mean that it would fail to breed, though submission is still its best bet because it might manage to find another place to live where it can start with a better social position. When the outcast is not fortunate a real curb is placed on the total breeding effort and another mechanism fashioned by natural selection to promote successful reproduction has had the side effect of putting a ceiling on a population.

It must be nearly impossible to indulge in a social life that does not have a spatial component. Consider, for instance, some recent discoveries about the social life of the white rhinoceros in a game park in Zululand. Norman Owen-Smith followed the rhinos about as Howard once followed birds, with results just as interesting. One of his first observations is that adult males spread out in a pattern that seems quite familiar to someone knowing about the territories of song birds. There is about two square kilometers for each of these males, and Owen-Smith was able to determine the boundaries round each male by watching his comings and goings. The bull rhinoceros does not sing to advertise himself, but defecates in piles, and kicks his dung with appropriate rituals of scraping feet and rubbing horn. He also sprays bursts of smelly drops from his penis as he goes about his daily rounds. The bull thus marks out a "do not come near me" space even as a cock bird gives warning with his twitter. When bulls meet at the corners of these spaces, the confrontations are completely reminiscent of the meetings of territorial birds. The bulls meet and touch, horn to horn, in what Owen-Smith describes as "A tense but silent confrontation." They back off again and strop their horns on the ground, then re-advance and tap again, head to head. Finally they back away from the place of meeting, each to his familiar ground. When one of a pair started the confrontation in what the naturalist had already determined to be the other's space, the errant bull backed up all the way home before the two bulls went on with their private lives.

This sounds much like Howard's territorial song birds at work, so much so that the modern naturalists who watch rhinoceri call the behavior by the same name, "territorial." There is even a floating population of non-territorial adult male rhinoceri, about a third of the total in the Zululand park, which may well be compared with the surplus of male birds demonstrated by the Maine Gunners. But there the similarities end. Female rhinoceri, juveniles, and even adult males who accept certain conditions wander almost at will across the "territories" of those who have them.

A territorial, male, white rhinoceros lets other males in, provided they accept a socially inferior status. In his territory the other male can live peacefully enough if it refrains from sex, never shows awareness that a passing cow is close to oestrus, watches impassively as the territorial male mates. Nor is the "territory" a homeland for a couple, a place for bringing up babies, or a private food supply. Females with calves, sociable pairs of male youths, all wander across the land, feeding here in the territory of one dominant male, there in that of another, in peaceful amity, seldom being molested. When a modern ecologist looks at the territorial behavior of song birds, he thinks particularly that the payoff is in food for the young. But we cannot conclude this when pondering the territorial doings of the white rhinoceros. Apparently we are seeing the resolution in two-dimensional space of different advantages accruing from behavior that says "keep away from me."

The clue to this territoriality of the white rhinoceros probably lies in the peck-order of chickens. We are observing a dominance hierarchy resolved in space, and the payoff to the owner of a territory lies in his freedom to copulate. The dominant male will let all classes of rhinoceros cross his territory except two: other sexually active bulls and cows that are approaching oestrus. The bulls he drives away but a hot cow he keeps, gently blocking her way when she tries to leave, making a squealing sound that is doubtless telling her something. She stays with him until she is ready and he has mated; then she is free to go. Territoriality in this species is thus a mechanism which resolves the social precedence in mating.

The payoff in fitness to the dominant male is obvious enough, as is the payoff to each bull of a pair who backs away from one of those tense and silent confrontations. More genes will be left to posterity by the rhino who backs down to wait for the sole services of a cow in his own space than by the rhino who spends his time fighting in somebody else's space. But there is a payoff in fitness for the subordinate bulls too. They are likely to be the younger males who may expect to inherit a "territory" as the old soldiers fade away and as they themselves have learned more. Owen-Smith reports that the taking over of a territory by another bull has been seen, after which event the old owner lives on, but stops his spraying and gradually ceases to kick his dung or strop his horn. Likewise subordinate bulls have been known to find a territory elsewhere. Cows benefit from the successful mating, which takes place in peace. Cows and juveniles wander and feed at will, finding the advantage of going to where food is most plentiful. All individuals thus derive fitness from a pattern of behavior, the genes for which they all carry, and the pattern has been preserved by natural selection because of this.

Another working out of social imperatives in space is illustrated by the separation of mountain lions described in Chapter Fourteen. Hornocker mapped the country traveled by each of several lions in winter, tracking the lions through the snow, running them down with dogs, shooting them with tranquilizer darts and identifying the lion that patrolled each piece of country. Hornocker found persuasive evidence that the lions did not defend their countries, neither by ritual display nor by fighting. He found the tracks of a prime male approaching the place where a juvenile was eating a kill in the corner of what was usually the prime male's country. The tracks showed that the older male had turned away before reaching the kill, suggesting that he had let the youngster alone with his food. Hornocker concluded that mountain lions avoid each other in winter, each keeping to himself so that he might hunt where the deer have not been disturbed by another lion. All lions win fitness from this behavior because it is in the interest of each of a pair to turn aside and hunt alone. This behavior is very different from that which yields food for babies in territorial song birds, or sexual peace for the white rhinoceros, but it is stamped out in space nevertheless. Hornocker called what he had found "Winter territoriality in mountain lions."

There is now a very large literature about "territoriality" in all manner of animals, in fish and insects as well as birds and mammals. It seems to be true that all forms of behavior with a component of "keep away from me" result in separations in two-dimensional space. We know of fish that have nesting territories like the song birds, where the male is gaudy and irascible like a robin in the spring, and of fish that have much more complicated arrays of feeding stations, particularly in places such as coral reefs. The rhinoceros pattern of a territory for dominant males that is essentially a sex station is so widespread among ungulates as to be nearly a general rule. And we know of many instances where "keep away from me" is applied by a bird or even an insect when the advantage of the behavior is clearly that it leaves the assertive one alone in a concentration of food.

All of these patterns of behavior involve the occupancy of space, usually alone but sometimes sharing with others who apply a rule to the non-initiated of "keep away from US." We call these various spaces "territories," because this is the easiest thing to do with our people-centered language, but this does not mean that the behavior really turns on the ownership of real estate. It is better to think of property as a consequence of behavior than as a cause. Not only does this have the advantage of directing our minds to finding out the real selective advantage of some of the intriguing things that animals do but it also warns us from making spurious comparisons between the "territories" of animals and the human propensity to aggression. Animal territories are seldom even analogous to human property rights, let alone the manifestation of similar "drives." The more extreme comparisons between the territorial extent of nation states and the spacing out of animals, of course, represent no more than a jejune falling into a semantic trap.

# Chapter Sixteen. Why There Are So Many Species

THERE are apparently very many more species living than there are obvious patches in the physical mosaic of the earth's surface; thus explaining this remarkable diversity of life becomes central to any attempt to understand the workings of natural ecosystems. Why so many plants and animals? Why not more? Our studies of the effects of crowding, of hunting, and of the changing communities through successions have shown us where the answers lie.

An ecologist's idea of a species is shown best in our exclusion principle, with its implication that for an individual living well within the niche of its kind the need to compete with other kinds is kept within bounds. It tell; us that all the kinds of grass and insect in a close-packec pasture must be going about the business of life in way; that do not press too heavily on their neighbors. This is not intuitively easy to accept. Grasses in a pasture seem manifestly to be struggling with each other for the right to live; they squeeze against each other, clawing upwards toward the light. And yet it is the ecologist's view that these struggles have been strongly muted and that there is really much peaceful coexistence in the pasture. It is necessary, therefore, for an ecologist to explain just how a place like a simple, open field can be divided up into so many non-competing niches. Furthermore, we must show how each of these many niches can be found by the process of natural selection in the first place, and how they can be fitted together so that there is coexistence in a well-filled space.

"Geographic isolation," or at the very least geographic separation along a gradient, is still the essence of our answer to the second problem of how the species are made and the niches found. Darwin relied on this "geographic isolation" in his original statement of the theory of evolution, noting that animals living in places far apart would be living in different local circumstance so that their "characters" might be expected to diverge. A whole section in *On the Origin of Species* talks of the "divergence of character." The idea is simple enough: different physical circumstances need different adaptations. But an ecologist's view of species lets us take the argument further than could Darwin.

It is common sense to imagine that two populations of the same kind of animal living far apart will differ in minor ways. The food or the weather in the different countries will not be the same, slightly different traits will give the advantage in each of the two places, and more babies will be reared by those animals with the genes for these local traits. Yet the two populations would still be of the same kind of animal. They could easily interbreed and mix their genes once more. Naturalists talk of local "varieties" and "types" of the same species, which will freely interbreed, and much of the breeding of domestic stock is based on crossing individuals from such local varieties. When Darwin wondered how separate species could arise, he had to guess that local varieties that had diverged far apart would somehow fail to breed when they met again. But we can show what actually happens in these fateful meetings. The Gause principle of competitive exclusion provides the clue.

If the two diverging populations had begun to get their food in significantly different ways, they would be tending to live in different niches. This is permissible as long as the two populations are kept apart, but if some accident of history lets them migrate together again, then there would be two groups of somewhat different niche-requirements trying to live together in the same circumstance, which is to say in the same niche. This cannot be; the exclusion principle tells us that this coexistence is impossible. One way out of the dilemma is for selection to force uniformity, by suppressing the deviant individuals so that only the original species remains. This must often happen. But there is an alternative way out, which is also often taken. This is to keep the two varieties so "apart" in the ecological sense that serious competition is avoided altogether. Selection for the most distinctly different animals in each population would accomplish this.

When two populations have diverged markedly, there may well be a few individuals in each who have developed the new traits or habits to so marked a degree that they can live together without competing. These few would then be favored by natural selection while the in-between individuals would continue to stray into each other's preserves. The extreme varieties would be safe from competition and would live on. The animals of the middle-ground would waste their energies in competing, leave few descendants, and be eliminated. The end result would be two, distinct, non-competing, coexisting species forged from the most extreme varieties available. We say that Darwin's "*divergence* of character" is supplemented by an actual "*displacement* of character" when the diverging forms meet. Natural selection is expected to force the characters apart so that competition is avoided.

We have found a number of patterns in nature that indicate we are on the right lines with our idea of "character displacement," the best perhaps being that of nuthatches in Asia. Nuthatches are entertaining little birds which get their living by running up and down the trunks of trees and picking out insects from crevices in the bark. They have short, stubby tails and oversized feet, and they often hold on upside-down as they peck—altogether comic and attractive birds, much prized as visitors to feeding tables. Most of them have a dark-colored stripe running through the eye, from the back of the head to the back of the beak, and this eye-stripe is almost certainly a species-recognition mark that helps them identify the proper mate.

In central Asia there is one common species of nuthatch, and in Greece and Asia Minor there is another. Both have long been recognized as good species by museum taxonomists, but they have presented "difficulties" all the same. These difficulties particularly concern the eye-stripes, for many individuals in Greece have eye-stripes of quite the same color and shape as many individuals in central Asia.

Somewhere in Iran, however, the population of central Asia to the east floods together with the population of the Greek species in a zone of overlap. In this zone of overlap, there is never any difficulty about telling the birds apart. In particular, one species has a vestigial eye-stripe whereas the other has a great thick black one. We say that natural selection has chosen the most extremely different individuals in the zone of overlap as the ones to rear babies. The shape of the eye-stripe must be a token for some other change, almost certainly in feeding habits, and only the non-competing strains can live together. The two species of Asiatic nuthatch are kept separate by "character displacement" where they live together in Iran.

Nuthatch populations wandering the vast expanses of Asia are but one tiny example of what must be a general and never-ending process. In each small patch of the terrestrial mosaic there is divergence of character as local varieties are made to adapt to local circumstance. If all the local populations were then left alone, we might expect to find a continual blending of types as we went about the earth. But the local populations are not left alone. Continual turmoil in the climate presses an endless ebb and flow onto the fates of populations, and the biological events of life itself also moves them around as species undertake their random dispersals. Varieties that have evolved apart are brought together, character displacement preserves the more extreme forms, and competitive exclusion removes the holders of the middle ground. Each separation and recombination of populations, therefore, may end with two species living side by side where there was one different species before.

We thus have a mechanism that continually forges new species and that lets us understand how so many distinctly different niches can be found. It answers the second part of our problem. But knowing how species are made does not explain how so very many of them can go on living together. We say that all those grasses in a pasture live side by side because speciation through character displacement has found ways in which they can grow in the same field with minimal competition, but this does not tell us what these different ways can be. The field is flat. The grasses touch one another. They use the same water, rely on the common reservoir of nutrients, endure the same seasons, and experience very nearly the same accidents. How, then, can they be living in such distinctly different ways that they avoid competing?

This problem of the pasture grasses is but one example of a common theme. It applies to tropical forests, where there may be a hundred kinds of tree in an acre; to the many kinds of tiny planktonic plants that live together in lakes and oceans, where the problem has become known as "the paradox of the plankton"; and it applies to all the much more diverse arrays of animals. We can understand that these local inhabitants can live together provided they do it without competing. Now we must show how there can be so many ways of not competing.

We can begin to get an answer by supposing that, for any community of land plants, there are irregularities in the soil. Any farmer or gardener will readily assert that soil irregularities do exist, and all we seem to have to do is to postulate a micro-mosaic of sufficient complexity to account for the complexity of the species list. There might be a grass-that-likes-a-patch-of-high-silicon and a grass-that-likes-a-patch-of-low-molybdenum and so on. The whole periodic table of the elements can be played with; this can be multiplied with degrees of wetness and dryness, by slope or exposure, and by some "biotic factor" of unknown nature and potency. There is no doubt that this sort of carve-up of the land does happen, and we *know* that different plants have different physical and chemical requirements. Some part of the diversity of plants can surely be explained in this way. But a hundred kinds of tree in a lowland rain forest? Can a patch of Amazonian mud really be a mosaic of a hundred distinct chemical brews? And all those grasses in a pasture; is there underneath them so beautiful a fine-grained pattern on the physical land? The hypothesis lacks conviction.

The chemical mosaic hypothesis has seemed in the past to be out of the question to explain the paradox of the plankton. Ever since Evelyn Hutchinson first pointed out the paradox fifteen years ago it has seemed "obvious" that the tiny plants were so stirred together in the chemical soup of lake or sea that each had the same chemical array to choose from as every other. Yet a recent paper by Richard Petersen shows that chemical variety can be reflected in species variety all the same. It is theoretically possible for different species of plant to be limited by different nutrients, so that the competitive power of the species wanes as the nutrient it needs most, and collects most efficiently, is used up. With all the competitive powers being set by different nutrients there is a separate curb for each species that would let a population of each coexist up to the limit of its own curb. The argument is complicated, theoretical and backed only by computer simulation, but it does show that chemical diversity can be read by natural selection to yield some species diversity.

But there are other ways of getting at the paradox of the plankton than relying on chemical limits for each species. They are based on the fact that conditions in a lake are constantly changing with time so that first one species blooms, and then another; they might compete but never for long enough for a species to be eliminated. Being tiny, the plants are short-lived; and being short-lived they can occupy the water surface in turn throughout the year. In lakes and oceans alike there are plants of the spring blooms, plants of the stratified times of summer when the warm water floats on top, and plants of the fall overturn, when the autumn winds mix the water once more. Each kind has a resting stage in a spore that sinks to the lake mud or that drifts the oceans in submerged currents. These plants separate their lives in time, instead of space, but this achieves the common goal of avoiding competition.

Having noticed that plants of the plankton partly keep themselves separate by using the water in turn, we reflect that there are similar opportunities for land plants, too. We see them do it in the plant successions, as the pioneer opportunists are gradually replaced by more equilibrium plants. At any instant in a plant succession (say an instant of a year or two) we seem to see plants living side by side in permanent array. But we know that many of them have only a tenuous hold on their living space, because some plants are always being squeezed out, and their successors are pressing their way in. So part of the paradox of the many kinds of plant in an open field may result from the fact that we are looking at a single frame in a process of continual replacement. There can be little doubt that some of the diversity in plant species results because opportunist and equilibrium plants operate on different time scales; they take the land in turn, each when the competition for its sort of life is least.

Such arguments encourage us to think we are working on the right lines. The physical world is diverse in both space and time, and we can imagine natural selection preserving those varieties that were fitted to different permutations of this physical diverseness as it worked its character displacements. The process would lead to many kinds of plant and, therefore, to many more kinds of animal—e.g., shoot-living animals, root-boring animals, and so on. We can see our way to accepting much diversity as being the natural outcome of these mechanisms. But the diversity we really have? Eight thousand species of birds, perhaps; but one hundred thousand species of vascular plants? And between one and three million species of insect? It is hard to believe that any permutation of an earthly mosaic can provide millions of distinct living spaces.

There is, however, a more potent force to keep the species apart, one that lets them live in common habitats without ruinous competition. This force is the activities of hunting animals, both the hunters of other animals and those hunters of plants that we call herbivores.

Farm animals such as cows and sheep are determined hunters of plants. They are not indiscriminate mowing-machines but animals of very definite taste, as any farmer knows. They do not want just any greenery; they want the things they like to eat. There are some very nice data from the celebrated sheep-pastures on the hillsides of Wales that show just what the ecological consequences of these cultivated tastes in a herbivore can be.

J. L. Harper used the records of many years collected by agriculture experts in Welsh universities to reconstruct what happened when sheep were grazed on different pastures. If a pasture is filled with the kinds of grass that sheep like to eat and then overstocked, two related things will happen. One is that the pasture will be ruined. The other is that the ruined pasture will have many more kinds of plant in it than the original, fine pasture. As too many sheep eat down, and kill all the plants they like to eat, they make room for the unpalatable plants of a ruined pasture to move in. There may be many kinds of these. Their numbers, when added to the survivors of the palatable populations, represent increased diversity in the field.

But if an old wild pasture, long used by a few sheep, is overstocked, a rather different pair of things happens. The first is the same; the pasture is ruined. But the second is strikingly different; it is that the ruined pasture this time has *less* kinds of plant in it than the original pasture. What the sheep of the heavy stock have done this time is to hunt down every one of the relatively few kinds of plant they do like out of all the array presented to them so that they kill all the palatable plants. At the end they leave behind just the lesser number of kinds they do not care for.

It is obvious that sheep on a wild hillside will keep down the populations of the plants they like to eat, thus making room for other plants to grow. These other plants, however, may be liked by a cow or a reindeer, which would, if present, keep their populations down in turn. This makes room for another set of plants, which will provide an opportunity for another grazing animal of equally fastidious but different taste. After a long evolutionary process we can imagine something like the game herds of Africa, where there are two hundred kinds of ungulate grazing grasslands and plains rich in plant species. Robert Whittaker tells me that similar richness is present in Israeli pastures that have been drastically over-grazed by goats, camels, sheep, cattle, and donkeys. These now have over one hundred vascular plant species per tenth hectare.

Yet most herbivores are not large mammals but insects, and these are more numerous and even more pressing in their attacks. Caterpillars strip the leaves from whole trees, beetles can drill every acorn or apple in a crop to deposit an egg in it. I discussed in the succession chapter how seed-hunting insects might determine the climax state in forests, whether it shall be a few dominant tree species as we know in the north or a brilliant array of species as in the lowland tropics. All plants must be subject to continual decimations from these attacks. If they are decimated, there will be room for new kinds of plants to evolve. Here, then, is an explanation for the rich species list of a pasture. The different kinds of grass are in the business of avoiding different hunters.

There is very good circumstantial evidence that all plants live their lives under persistent attack by herbivores, and that this attack may be mortal in some seasons and at some stages in the life history of the individual. Most persuasive are the chemical defenses of the plants. Plants contain many chemicals that make them smell or taste strange, or that are actually poisonous to many animals. The presence of these chemicals makes sense if they ensure that the plant is avoided by a potential persecutor, because the trait of making the chemical will then be preserved by natural selection. The niche of a plant with a nasty-tasting chemical may be said to be partly characterized as not-being-eaten-by-herbivores-X-Y-and-Z (which do not like it). This kind of plant then avoids competing with another whose chemicals are different and which repels a different set of herbivores.

Some plants have mechanical devices to defend themselves, such as thorns and spines. They can also be defended, as species though not as individuals, by being able to disperse their seeds far and wide. One obvious advantage of a wide dispersal of seeds is that the young plant may be able to grow big and strong in an odd corner where it may be overlooked by grazing hunters working a methodical system of search and destroy, as does a sheep cropping its way across a pasture.

Every specialized defense provided for a kind of plant by natural selection gives an opportunity for specialized attack. If a plant is poisonous to all the animals around it, the time will surely come when a strain of one of the local herbivores will appear that is immune to the poison, and this strain will be favored by natural selection. The results of this process are evident in what we know of the habits of very many insects. Caterpillars of butterflies and moths, for instance, can generally eat only one or two kinds of food plant. A similar thing happens when a plant is provided with thorns (at great cost in calories) against grazing animals. Acacia bushes in Africa meet their match in giraffes, which can delicately snip the leaves from between the thorns.

Every defense sets the evolutionary stage for a new attack; which gives the opportunity for a new defense. Every new kind of animal that evolves to hunt specialized plant food, provides, in turn, an opportunity for a new kind of carnivorous animal to hunt it. And this happens at every link in the food chains until the second law of thermodynamics says the available energy becomes a trickle and calls a halt to the length of the chains (Chapter Three). But near the base of the food chains, where herbivores hunt plants, the process is without end, a continuous open-ended contract with increasing diversity. This is the crux of our explanation of why there can be so many different kinds of plants and animals. Ecologists now refer to the working of this process as "the cropping principle."

But one question remains. Why are there not even more kinds of plant and animal? Why only three million insect species and a miserable tenth-million kinds of plant? We have just discovered an open-ended process for the making of new species that has been operating for hundreds of millions of years. Are the numbers we now have all that could be made in that time? Are these numbers just accidents, or is the earth full? And if so, what does "full" mean? Ecologists are still inclined to argue about these things, but it does look as if we might have the general answer to these questions, all the same. The answer comes from twin reflections: that species go extinct even while others are being created, and that different parts of the earth have quite different numbers of animals and plants living in them.

The northern lands beyond the arctic treeline are well-vegetated in the sense that the ground is almost completely covered with plants, but relatively few kinds of plant make up this vegetation mat of the tundra. There are more kinds of plant in the boreal forest to the south, more still in the temperate deciduous forest lands and the prairies, and very many more still in the tropical regions. The same sort of thing is true for animals; a few kinds in the extreme north and progressively more with every degree of latitude south down to the equator. There is, indeed, a general cline of diversity running from the poles to the equator, with more snakes, more insects, more mammals, more ferns, more grasses, more everything in the tropical lands. To understand why the earth has the numbers of species which it has, we will have to explain this gradient of diversity from north to south.

The answer obviously has something to do with the rigors of climate. The arctic regions are cold, water freezes there, and the night is six months long. This in itself is good enough to explain the absence of plant and animal designs that cannot endure these things: of frogs that must be kept moist, of succulent plants that burst if they freeze, of trees that cannot balance their heat budgets (Chapter Five). But it is not enough, by itself, to explain the paucity of insects and grassy herbs. Some kinds of insect and some kinds of herb live on the arctic tundras, even thriving there and attaining very high populations. If some can, why should not more kinds share the flat arctic spaces as they share the flat meadow fields of southern places where there is less weather to worry about?

The best answer we can give is that life in the north is accident-prone. From the equator northward, the weather gets not only systematically more hostile but also unpredictably so. The unseasonable flood, frost, or drought becomes ever more likely as the warm girdle of the earth is left behind. This means that the chances for catastrophic accident also increase nearer a pole. And this increases the chance of being made extinct.

Furthermore, all animals and plants of extremely seasonal places have life-strategies that reflect the world of boom and bust in which they live. They rest in winter, they germinate in spring, they rush their growing efforts through in the brief weeks of uncertain summer, they try to be inert again before the next winter makes its unpredictable beginning. These very life-strategies, essentially opportunist, impose histories of fluctuating numbers on populations of seasonal lands. Boom and bust in the physical environment means boom and bust for populations. And it is on populations like these that the random hostilities of unseasonable events must fall. If a population is hit by catastrophe when it is already in one of its low states, there is a very real chance that it will go extinct.

The best explanation we can offer for the declining numbers of species from the equator northwards is that it reflects a gradient of chances of being made extinct. If new species are forged everywhere at the same rate, whereas old species become extinct faster and faster as one moves northwards, then a gradient of diversity should result.

Striking support for this interpretation has recently come from the work of an oceanographer, Howard Sanders. Sanders revealed another gradient of diversity across the face of the earth, and one that had not been expected. This gradient runs out to sea from the coast of a continent, over the continental shelf, and down onto the floors of the abyssal plain in deep water. It is a gradient of diversity in the animals that live in the bottom mud. And it looks as if it runs the wrong way, because there are more kinds of animal living in the inky blackness and everlasting cold of the deep sea floor than there are in the warm sunlit bottoms near the coast.

Sanders showed that in the sunny, productive, coastal waters, where we should have supposed it was "nice" to live, there were in fact few kinds of bottom-living animal. Down on the continental shelves, a more gloomy place and somewhat removed from the living action and food supplies of the surface, there were distinctly more species. At the bottom of the sea, where the pickings must be very lean indeed, there were the most species of all. This made sense, however, if the numbers of species living in a place are set by the chances for accidents that could drive some kinds extinct. The cold, dark ocean floors have no weather. Their temperature never changes. Their darkness is always the same. The lean pickings of falling debris probably do not fluctuate much. There has been no news to report for millions of years on end. Therefore, says Sanders, the chances of going extinct are very small; speciation goes on, and the diversity tends to rise. But the supposed "nice" productive bottoms of coastal waters have a great deal of weather. The local species, like those of the arctic, have opportunist life-strategies designed to cope with boom and bust. Their numbers fluctuate. They can be hit by freak accidents. Although the place appears "nice," life is fraught with the possibility for accident.

Our general answer to the question of why there are not more species than there are is that there have been more, but they went extinct. This lets us give a general answer to the whole grand question of why there are so many kinds of plant and animal.

Geographical isolation leads to divergence of character, and this goes on constantly. As diverse populations merge, natural selection preserves the most distinct individuals that can live together without competing. This is the crucial step in the process of evolution, and we call it "character displacement." Very many of the choices made by natural selection are of individuals that have new ways of hunting, or new ways of avoiding hunters. This is what we mean when we talk of the "cropping principle." New species are being made constantly all over the earth by this method, but they are also being removed by random hostilities in the environment that drive some to extinction. The chance of becoming extinct is higher in some places than in others, so that some places hold on to fewer species than do others. But the average numbers of kinds of animal and plant over the whole earth are set by a balance between the rate at which new forms are made and the rate at which the old ones become extinct.

# Chapter Seventeen. The Stability in Nature

IT is said that where many different kinds of plants and animals live together there will be a better balance than where there are only a few kinds. This is to say that complexity leads to stability. Ecologists have been saying this in the recent past, quite loudly, and they have been quoted by those concerned about the human impact on our planet. But ecologists are now tending to eat their words. What follows are both the words and the eating of them.

The stability argument can best be understood in an extreme example. If only two kinds of animal exist on an island, say foxes and the rabbits they eat, then the future of both kinds looks highly uncertain. If some accident killed off many of the rabbits, this would be very unfortunate for the foxes, most of whom would starve. The few rabbit survivors of the physical catastrophe would also be in an extremely precarious state because they would be hunted down by the relatively numerous and desperate foxes. But if the natural accident happened to the foxes, then the numbers of the rabbits might get out of hand until more foxes had been bred to eat them up, by which time there might be too many foxes, and so on. The fox-rabbit system would be dangerously unstable.

But if instead of two kinds of animal there were as many as ten different kinds of rodent on the island living with the foxes, say several kinds each of rats, mice, and voles. And if there were two or three other kinds of flesh-eater as well, say cats and weasels in addition to the foxes, then in this well-populated island a catastrophe to any one kind of animal would not matter very much. If the rabbits on this island suffered catastrophic loss, the predators would still be safe, being able to feed on the other nine kinds of rodent. The rabbits might also survive their catastrophe because it might not be worth any of the predators' time and effort to specialize in hunting rare rabbits, and the few remaining rabbits might be left alone to pursue their exuberant breeding policy to make good the loss. Similarly, if the foxes suffered accident, there would be no wild fluctuation in rodent numbers because the cats and weasels would be there to carry on the hunting. They might even expand their appetites to eat the rodents left by the missing foxes. Life on this hypothetical well-populated island, therefore, should be stable and safe from extinction.

This is the essential part of the complexity-stability theory, a straightforward idea whose only unusual or tricky aspect is that it is the complex that is stable. The theory has deep appeal to naturalists for it fits the intuitive idea of complex nature working well. This feeling was there when we looked at the great plant formations and saw them as entities, with territories set aside to each. Then the search for plant societies as real communities of species interacting to preserve the common order revealed the same thoughts. So did the idea of successional communities being but subordinate stages in the building of a climax formation. And the idea is very strongly present in all thoughts of a balance in nature set by predators that are supposed to control the numbers of everything; the spiders that are "good" because they kill "flies" and the wolves that are "bad" because they kill "game."

But the theory only became important in modern ecology when claims appeared that it had a firm basis in mathematics, and satisfyingly erudite mathematics at that. The erudition may have been the snare in which we were caught, for the mathematics never said what ecologists came to think it said. Telephone engineers of the Bell System's laboratories did the math. They were interested, needless to say, in complicated networks of channels down which messages flowed, and the mathematics they devised is called "information theory." The theory provides a measure of the diversity of channels in a network, called the "Shannon-Wiener information measure" after its authors. The math also states the relationship between this measure and the capacity of an information channel. If this strikes the casual reader as apparently having little to do with biology, this shows the reader's good sense.

To get from the Bell System to an ecosystem we first use the Shannon-Wiener measure to describe the diversity of species in a biological community and then we indulge in risky analogy as we compare one system with the other, intuitively. The first stage in this process, the use of the measure, seems to be reasonable and useful in that it speaks to a very real difficulty we have in describing biological systems. The measure helps with our perennial problem of the common species and the rare, particularly the proper description of commonness and rarity. It is a fairly easy matter to list all the species in a community and to compare the species lists of two communities in the ways of the plant sociologists. But what if two communities are made of the same species but these appear in different proportions? Obviously the communities, and the ecosystems that support them, will be different. We say that the two have the same "species richness" but different "species diversity."

Those who love the English language will notice that we have given our own special meaning to "diversity."

We use the Shannon-Wiener formula to measure diversity in collections of species because it allows us to collapse estimates of species richness and species commonness into a single statement. There is a large ecological literature on when and how to do this and ecology has benefited from the practice. But it is from here that the errors begin, because a measure of species diversity must also be a measure of complexity. And the original information theory gave both a measure of the array of alternative channels (diversity) and of the capacity of a channel for the flow of information, which gave the stability of the flow. If the measure describes both complexity and stability, which it does for the system of message channels, it is very tempting to think that Shannon-Wiener measures both complexity and stability when applied to biological systems also. And so there we have the snare. We use a measure from another discipline to describe the diversity of our ecosystem, and find that it does do so in a general way. But then we notice that the measure also describes stability in the phenomena of that other discipline and we are tempted to make the claim that the measure describes stability for our phenomena too. But the phenoma are not the same. We get from Bell System to the ecosystem by analogy only.

Ecologists in the late 1950s suddenly became aware that the telephone men had produced a body of theory that seemed directly to relate complexity to stability in physical systems. Ecologists were thinking "systems," and were in fact actively teaching their students that "the ecosystem" was the unit to study. And here were systems theorists with elegant mathematics purporting to show that complex systems (ecosystems?) should be stable. It was no more than what an ecologist had always expected. Those richly diverse communities that botanists had once called "formations" or "associations," and which Tansley had said were to be looked upon as parts of "ecosystems," were able to persist because their complexity gave them stability.

The natural history literature contains many anecdotes that support this view. On the one hand we have the rain forests of the equatorial basins, biologically rich places with more species than anywhere else on earth. These were communities of immense complexity, and we thought of them as unchanging, timeless ecosystems that had endured for whole epochs in uneventful sameness. On the other hand were the arctic tundras, with few species, where the records of fur traders told us of violent oscillations in the numbers of animals, and from which came stories of lemmings taking intermittent marches to the sea. The complex place was stable and the simple place unstable; just as the theory predicted.

More potent still for the success of the theory was its apparent usefulness in describing the difficulties known to farmers. Western agriculture works by clearing the wild complexity away and substituting a single crop. Where there were formerly deciduous forests or prairies, with their complex lists of species, we substituted monoculture; one immensely common plant with a few hangers-on in the form of weeds. This is creating extreme simplicity where before the system was complex. Information theory predicts that the new ecosystems built by the farmer should be unstable and, lo, the fields of agriculture are afflicted with plagues of weeds and plagues of pests. It seemed like an ecological judgment.

But a closer look at these anecdotes causes disquiet. The instability of arctic animal populations seems clearly to have something to do with a highly unstable climate.

Indeed, we call upon the vagaries of the arctic weather for our best explanation of why the area is depauperate, saying that species go extinct in the arctic so quickly that a large species list cannot collect (Chapter Sixteen). And we explain the rich species list of the equatorial forests as being due to the fact that there is so stable a climate in the equatorial lowlands that extinction is rare, letting more and more species accumulate. This introduces a dilemma of priority; does a large species list promote stable life? Or does stable living in a place of stable climate promote a large species list?

Adding to these doubts was the realization that we were not really sure that life in the equatorial forests was stable. We had very few data because very few modern biologists have lived there. Western civilization, and its biologists, are products of a narrow band of latitude circling the northern hemisphere, half-way up toward the pole. We have more than a passing interest in what goes on to the north because we hunt arctic animals for their fur. People of our northern outposts have reported what they have seen, and when they have seen something unusual they have reported it all the more vehemently. But we have much less news of the rain forests, nor have we had a commerical interest in the systematic collection of small tropical animals. If there have been plagues of mice or monkeys along the Zaire River or in Borneo, we have had no resident scholars there to write to the *Times* about them.

Now things are beginning to change. Recently a scientist with long experience of the arctic settled in Panama to work, and wrote to a scientific journal saying that he had seen as many rodent plagues in four years in Panama as he had during his longer living in the arctic before. I recently flew low over the rain forest in Ecuador and saw scattered trees that had lost all their leaves, perhaps because of a population event in the caterpillars that feed on them. Many years ago in Nigeria I saw the same sort of thing from the ground. One species of tree in the local rain forest was suddenly easy to spot because it was without leaves. A plague of caterpillars had totally defoliated it.

These stories are only anecdotes. But so are the accounts of fluctuating numbers in the north. Neither is a measure of stability; both are merely accounts of fluctuations in the numbers of individual species. The point is that we realize that we are probably going to be able to match descriptions of population events in the depauperate north with descriptions of similar events in equatorial places, where large arrays of species live. We cannot rely on comparisons between latitudes to support complexity-stability theory, quite apart from the difficulty of prime causes introduced by different climates.

The arguments based on agriculture, when examined closely, are even weaker. In essence they say that very simple systems such as the ones the farmer makes should not work at all. A field of monoculture could be likened to my first model of an island inhabited only by rabbits and foxes, with the crop playing the part of the rabbit and the farmer or his pests playing the part of the fox. The system should be wildly unstable, which means that agriculture should not work. But Western-style agriculture does work, very well indeed. The crops and the farmers both thrive, as they have done for the ten thousand years during which agricultural systems have become ever more simple. It is a triumph of stability.

There are troubles of the eggs-in-one-basket kind for farmers in practicing monoculture but these are not strictly relevant to the complexity-stability argument. When accidents happen to a monoculture crop they are likely to be catastrophic to local economies, but this is the consequence of not spreading the economic risk and does not speak to the fate of the crop species itself. It is probably true to say that the fortunes of crop plants have little to do with the systems properties of simple communities. If there is also no more instability in the north than in the tropics, which cannot be accounted for by the instability of weather, then there are no general biological observations that can be used to support the theory. It becomes no more than an echo of beliefs in natural organization held by the old naturalists who thought that there were self-organizing powers in plant societies or ecological succession.

Information theory itself is certainly valid. Systems that function through an array of intersecting pathways that provide alternative channels for the flow of information or energy do, indeed, become more stable the more crossroads there are. The colossal error behind the application of the theory to biology is in imagining that animals and plants in a food web act as the necessary crossroads.

Real animals and plants do not conduct themselves as channels for the transfer of that important form of "information" or energy called "food." They work to stop the food moving. Every individual of every species in the community is working its hardest to secure food and to prevent others taking it. The information theory description of a food web sees each individual as a channel at a crossroads through which food freely passes, but real individuals are in fact road-blocks through which food gets with difficulty. It is this fact that makes the model not only unreal, but absurd.

The ecosystem concept is beautiful, letting us express our understanding of how the doings of every kind of living and physical process in the habitat may affect the fortunes of all. With information theory, however, we stretch the systems analogy too far. It requires that animals and plants act in ways we know they do not. In particular, the theory relies heavily on the efficiency of predators, expecting them not only to control their prey in a very simplistic manner but to be catholic of taste so that they can turn their formidable mouths to whatever victims happen to be plentiful. But we know (Chapter Fourteen) that real predators do not work like this.

Most hunting animals are the small insects, like wasps and beetles, and these are highly programmed to hunt particular kinds of prey. They do not switch their attentions from target to target as the theory requires. Waging their guerrilla war of hide and seek through a tropical rain forest a kind of wasp and a kind of caterpillar are really as alone as the foxes and rabbits of my imaginary depauperate island. Their fortunes are not given stability by the presence of neighbors. They persist only by the logic of run and scatter, search and destroy. They would do the same in whatever community they lived.

For the herbivores, the hunters who eat plants, the reality is the same. Each specializes in its own kind of plant, or its own few kinds, so that the system of interchangeable channels required by information theory does not exist. Again this is most true for the small insect hunters, which are often totally dependent on a single plant species for their livelihood. But most of the complexity of species in the tropics is made up of insects and the plants they hunt. In real communities the animals and plants live much of their lives in isolation from their neighbors of other kinds. It is peaceful coexistence, as the exclusion principle tell us, not the constant death in skirmish that the information theory model requires.

Recently this biological theme has been taken up by mathematical ecologists. Hitherto we had been applying to ecosystems the mathematics appropriate to telephone networks, or simplified physical systems provided with free-flowing feedback loops. It has led us into great error. But now the first systems models are being made on assumptions that the units in the systems behave as we know animals behave, where the feedback between one event and another is resisted or delayed. For these models, there is no simple relationship between complexity of species list and stability in the lives of populations. Indeed, a common result is quite the reverse. In some of these models, if a complex "community" is perturbed, the result is not stability but reinforcement of the stress, a domino effect of increasing instability the more species there are. There is actually a resonance, with the original fluctuation being amplified as the shock travels through a complex community.

The claim that complex communities are more stable than simple communities, therefore, is invalid. It is an echo of the wishful thinking of naturalists, amplified by mathematics they did not understand. It has done mischief by distracting people from real problems. It has, for instance, been invoked in the controversy over the Alaska pipeline, in the claim that the arctic ecosystem is "fragile" (it is *simple* don't you see). But this is nonsense. The animals and plants of the arctic spend their whole lives and evolutionary experience struggling against adversities far mightier than any pipeline or road. Fluctuating numbers are normal conditions of many of their lives, all of them will outlast oil-hungry people. I happen to think that the Alaskan pipeline is a disaster to the American heritage, both for the aesthetic damage it does to the last wilderness and for the encouragement it gives to the continued misuse of fuel reserves. I wish very deeply it could have been stopped. But the argument that it is damaging a fragile ecosystem is false. Far more serious for the future is the trans-Amazonian highway, because this brings the shock of human activities to a rich variety of tropical species so unaccustomed to shock that many will disappear forever with the coming of the road.

Many species in an ecosystem do not, of themselves, lead to population stability. Stability of climate, on the other hand, leads to the collection of many species. This seems to be the essential truth of the matter.

But then what does cause the balance we see about us in nature? Many different things conspire to preserve the continuity of life, which is what we mean by balance, but central to them all is that fact that every species is equipped with a strategy for life that lets it persist. A climax tree of the forest has the strategy of holding ground, long life, and growing as a baby in the shade of mother. It takes generations, a hurricane, or a plague to displace climax trees, yet both hurricanes and plagues are rare. And among the true climax trees are patches where holes are being filled by succession, but change is slow even in these places. So generations of people see the same forests, even though they will not last forever.

Weed plants come and go, suffering numerous upheavals, but their opportunist strategies let them plant a new generation in some new spot as fast as the old is deposed from the parental site. The coming and going of the weeds is always with us, and we see in this continuity part of that general balance. Herbivores hunt down their plant prey and move on, but the piece of land they clear will immediately be taken by another plant because that land is unfailingly supplied by the energy of the sun. Plants and their persecutors go on with their endless game of musical chairs; and we see the result as balance.

Insect predators, and the other hosts of small hunters, pursue their games of search and kill, hide and seek, with their scattered, mobile quarry. This too tells us of persistence, we say "balance." The larger predators maintain themselves on the old and sick; they are long-lived and must survive the winter; they must therefore be rare, and will suffer deaths of privation if there come to be too many of them. This too contributes to the general balance, and comes closer to the tooth-and-claw model of balance set by quarrels about limited resource. So too do the activities of the big hunters when they kill the young of their prey, suppressing the population of their victims and thus limiting their own eventual supply of old and sick.

The birds and many other vertebrate animals also have complicated patterns of behavior, which are necessary for them to rear their babies and to survive hazard, and all these have population effects. The territorial habit, whether giving advantage through food, union of parents, or sex, always carries with it the possibility of setting an upper limit to numbers. So do all the hierarchical structures of social animals. None of these things has evolved to promote "balance" by restricting breeding, but all may tend to have that effect. These habits mean that many of the more conspicuous of the activities the naturalist sees are nearly the same every year. This gives us a feeling for a general balance in nature that would not have been so strong if we had fastened our minds on plant-eating insects or ichneumon wasps or spiders.

It remains true that the natural balance involves great destruction every year, because all species are breeding as hard as they can. But this natural destruction mostly falls on eggs or the young. Beetles drill nearly every acorn. Dandelion seeds floating under their parachutes mostly fall on stony ground. The hunting wasps get growing caterpillars. Yearling animals are the ones who fail in hard winters. And, when times are very hard through too much crowding, as can happen in nature and always happens in laboratory cages replete with food, it is eggs, embryos, and young that are starved, even as the old die untimely deaths. It is very hard to raise babies in the real world. It is the unmade or the unfinished animal and plant that succumbs. In a sense, nature favors abortion rather than later decimation by tooth and claw.

# Chapter Eighteen.  The People's Place

PEOPLE are animals that have learned to change their niches without changing their breeding strategy.

Most of our history has been passed in the ice ages, when the climates of all the world changed in accord with the advance and retreat of the glaciers that came pressing down from the north. The last, long period of peace our species knew was spent in the time of the last glacial advance, a time of longer duration than the ten or so thousand years that have passed since the ice went away. We lived then all over the world, in forests and savannahs of the tropics, in forests closer to the ice that were like the forests of temperate Europe and America now, and on the dry steppe-tundras that covered the plains of northern Europe, Russia, Siberia, and the lands now lost beneath the Bering Sea. We hunted and gathered, sometimes more of the one, sometimes more of the other.

When we hunted, we lived like tigers or wolves as the top predators of our day. Perhaps we often killed the old, sick, and young as the modern wolves do, but perhaps we killed more, for we were better armed than they. A spear with a stone point kills more swiftly than any tiger's teeth, and the attack plan forged by intelligence lays the more deadly trap. It may be that we took prime specimens of our prey as well as the feeble. Be that as it may, we still operated as the top predators of ice age time, leading an active energetic life at the ends of our food chain. We were accordingly rare, as lions and great white sharks are rare.

When we gathered, we lived like bears, collecting fruit, nuts, and caterpillars, and getting some meat from hunting on the side. We wandered in family units as bears often do, but we probably made a better thing out of it by rationing our resources, planning our journeys, and storing food for the leaner seasons. There was probably a division of labor in the family, with the women, made partly immobile by their children, spending much time on gathering while the men foraged further and hunted bigger things. But the way of life used the sorts of resources that bears use, and the people were accordingly as thin on the ground as are bears.

So we lived in niches that won food supplies comparable to those of tigers or bears, but there were already in our species niche many of the peculiarities that typify us today. We clothed ourselves, and rather well at that because there were people on arctic tundras during the last ice age, where clothing would be necessary to stay alive. Our ancestors in those distant peaceful times could make boots, mittens and parkas if they had to. We also lived in houses, whether these were just caves or huts made from ribs, branches, skins, and sods. We had enough of a social life to organize to hunt big game, to separate in the scavenging season, or to pass long winters in close quarters. We may already have started that long association with dogs that we still cannot shake. In all these ways, the people who got the food that bears or tigers get were already like us. It is for their life that we were fashioned by natural selection.

During long millennia of geographic isolation, natural selection began to fashion geographical races and varieties of people, some of which involved changes in the pigments of the skin. These varieties of people, however, were never very distinct. They involved no significant change in the important niche parameters like those that are the basis for character displacements when new species are made. When accidents of history and emigration brought any of the geographical races together again there was nothing for natural selection to choose between them, as it does for the nuthatches of Asia. The races were so essentially similar in all the things that mattered that they freely interbred and mingled their genes where the populations overlapped. Natural selection does not remove the holders of the middle ground in favor of the extremes where the human races meet because all are equally good at the people-niche.

We were rare, as all large fierce animals are rare, but, also like the rest of them, we had a breeding strategy that ensured that the maximum possible number of babies would be raised. We were made to follow the most efficient of alternative breeding strategies, the large-young gambit, and we were given the gambit in its most extreme form. Our young were at first completely helpless, and then had to be nurtured for an astonishing ten or fifteen years. Even then, they were not fully mature adults, but were at risk for more years still before being ready to be successful parents, able themselves to accept the onerous duties of raising more people.

This breeding strategy can be successful only if the cost-accounting is very well done. The ambitions of each couple must be carefully programmed so that they start with just the right-sized family for their local circumstance, because mistakes at the family-planning stage will be ruthlessly punished by natural selection. Any couple that is abstemious, having less children than they might have raised, will contribute less to the next gener-

ation than those with more ambition. I
dren, and children's children, were equa
their line would die out. So natural selec
allow our ancestors to have smaller famil
could.

But too large a family might be more dis
because the family's resources in a tough wi
be stretched so thin that all the children w
There must be a careful and accurate choice o
timum number of children. This is required
species that follows the large-young gambit (
Two) and pairs of people have always had to be abl
sess, rather precisely, the exact number of children
could afford. Our ancestors were helped a little
their choosing by brute animal needs, by feeling the
ness of their bellies and the fatness of their flesh. W
fed females came into breeding condition at young
ages than those that lacked protein, and it is likely tha
carrying a fetus to a successful live birth has always
been influenced by the well-being of the mother. These
mechanisms still work in deprived parts of the world.
But the people needed, and were provided with, better
systems than these for ensuring an apt family size.

People could use intelligence for this business of choosing the numbers in their families. The very long juvenile
apprenticeship that people serve argues strongly for ra-
tional choice because the pay-off in fitness for every in-
vestment in a child is delayed for twenty years. In
operating their Darwinian breeding strategy people
must always have been looking into the future. Sometimes their reasoning may have been clear and direct;
learning from the Joneses all of whose children starved
one winter because there was not enough food to go
round, and from the Robinsons who made it through
with their more modest brood. Sometimes people may

f their few chil-
lly abstemious,
tion would not
es than they

astrous still
nter might
ould die.
f the op-
f every
Chapter
e to as-
they
with
full-
ell-
er
t

he old folks told
han, intelligent
s. I am inclined
selective advan-
this elegant tun-
ligence may have
ossessors regulate
g effort.

every month and
that seems to know
about lovers and the
xual habits can be ex-
person-starts so that the
st have been with our re-
ate a breeding strategy that
imum number in these circum-
ires a mechanism for culling the
did was to let babies surplus to
s die. We call this practice "infan-
ow that the habit was once widespread.
gence that sees its necessity, it is a pecul-
habit. And it has the astounding result that
s the growth rate of a population. This is an
aps so novel that it needs to be said twice. In-
e is to be expected to *increase* the growth rate of
pulation, not to restrain the rise of numbers. Infan-
de is killing surplus babies for whom there would be
oo few resources in order that others might survive. It is
the mechanism of culling inherent in our sexual behavior
and our Darwinian breeding strategy. Infanticide confers fitness.

Yet infanticide consciously used in the regulation of
family may often have been less important than baby
death, and other reproductive restraints, passed on for
generations by another new trick that our kind long ago

introduced into the evolutionary gan
tion.

Human clans or tribes pass on habits f
tion to the next by conscious learning,
genes. When this ability was first acquired
thing wholly new. It meant that the succe
in the pure Darwinian sense of leaving th
vivors could be influenced not only by thei
by fate but also by what they had learned. P
in close-knit groups, listening to their own
people; and each group developed its own spec
doing things, what we call its "culture." If a gro
culture that was better suited to the rearing of
than a neighboring group, then people trained
successful culture would replace those trained i
other. In this way, cultural selection would lead to
sole occupation of a piece of country by people with
culture most suited to raising the maximum number
children. This successful culture would be particular
one that guides each couple in its choice of how man
children to try to rear, by helping them decide how
many children they can afford.

Nothing need be consciously intended in the successful trait; all that is required is that people do it because it
is expected of them. Any form of sexual taboo would
serve if it prevented the chosen family from being
dangerously large, so would sacrifice or infanticide,
whatever the intentions of those who perpetrated them.
It was enough that young couples should be sufficiently
influenced to do what the witch-doctors or old women
told them to do, whatever mystical purpose they imagined they were fulfilling. Provided the trait kept the family size to the optimum, so that each couple "chose" the
number of children it could really afford, it was possible
for cultural selection to preserve the trait.

People's Place

...e, cultural selec-
...om one genera-
... without using
...l, it was some-
...ss of families
...e most sur-
... genes and
...eople lived
...old wise
...ial way of
...p had a
...babies
...in the
...n the
...e the
...the
...of
...ly

...ometimes meant
...would always
...em dying from
...n and neglect.
...he milder acci-
..., time needed to
...hard for young-
...their lands (the *K*
...hittled away. For
...cted traits such as
...rituals would have
...adapted people very
...niche in their ice-age
...d thousand years their
...ay.

...s ago people learned to herd
...it, and invented agriculture.
...ood resource, because it denied
...ce at the game we had corralled, it
...mals when we wanted them, and it
...calories in the inefficient process of
...ulture increased our food supply still
...; with it went a plant diet that would even-
...y move down the Eltonian pyramids of our
...whole trophic level. We should henceforth be
...ap the rich energy supplies made available by
...ants at the bottom of our food chains. By control-
...which plants should grow we then had in our power
...e tool that would one day let us divert the primary
production of the whole earth to our sole use.

Herding and agriculture entailed the adoption of entirely new niches. For the first time an animal had
adopted a new niche without speciating. It was the most
momentous event in the history of life. It meant that one
kind of animal was now able to keep changing its habits

in ways that should take the food fr
would pay no cost for this in the loss o
cost always paid in changing a niche thr
People would take away the resources
the niches of other animals one by one, c
ing these resources to their own niche.

It is not intelligence itself that sets huma
other animals. People have been intell
hundred thousand years, but lived in their
place like the rest of the animals. They o
ecological rules of peaceful coexistence. Th
thing setting people apart from all other living
their ability to change their niche at will. It me
they are "Without the Law" as Kipling would ha
People knock out the other species as they make
changes. We first acquired this power only
thousand years ago.

Yet, when people made this momentous change, t
did not escape all the restrictions of their ancient nich
People were adapted to the life of the wandering ice-ag
foragers by physique, by temperament, and through
many subtle desires and patterns of behavior. These
things remained and do so today. Children continued to
invent games that would fit them for the adult tasks of
hunting, collecting, and preserving what had been won.
People still clothed themselves, lived in houses, and
held to our ancient alliance with dogs. They never lost
the conservatism that had once kept them safely within
the bounds of their old, rather rigid, niche. And they
held fast to our Darwinian breeding strategy.

Every couple continued to raise the number of children it thought it could afford, just as they had always
done, and they kept habits that let them estimate the
number of babies their circumstances could support.
Soon generations were reared who had known none of

People's Place

om others, and it
the old ways—a
ugh speciating.
that supported
onstantly add-
s apart from
gent for a
appointed
eyed the
crucial
things is
ns that
said.
their
nine
ey
e.
e

hen there were
to see them
evitably grew
ued for the re-
ial restraints of
d as the young
re children than
d families were
nd the extra food
egan to be turned
tionary way, even
food were entirely
an the geometric in-
t abundance.
ense, and settled. Food
ed and rationed or the large
could no longer be fed. The
t economy appeared, both the
es of dense human populations.
the food supply for large numbers
bad, hoarding, rationing, importing.
tells us a little of these early days. We
tes, who were still herdsmen living in
nadic populations, traveling to save their
om hunger because "there was corn in
The Egypt to which they journeyed was a
d state with magazines of grain.
the city-state there had to be governors and
erned—those who organized, and those who were
ontent to wait for the next meal. So now there were
many different niches lived in by individuals of this one
species. The lives of the organizers, be they merchants,
bureaucrats, or priests, were wide-ranging and required
many resources of the living space to sustain them.
These rulers had broad niches. But for the mass there

was a constricted way of life that needed little more of the living-space than the resources to grow food. These people of the mass had narrow niches.

People had invented wealth and poverty. For some the trick of changing the niche led to marvelously widened aspirations. They had leisure to invent the arts, cultivate their minds, and to plan ways to live in physical comfort, all of which came as a direct result of that curious trick of changing the way of life without speciating. And these lucky people need pay none of the cost of giving up the activities of their ice-age niche for which their bodily and mental mechanisms were made. They could go adventuring or could hunt. Thus were the large, compounded niches of the rich. But the poor could only eat, toil in ways for which they were not well adapted by natural selection, and breed.

A large niche requires more resources than a small one, whether these resources are space, food, energy, raw materials, or more subtle things. There cannot be so many of the large niches of wealth as there are of the small niches of poverty. Because of this, the early glut of babies that followed the new ability to change the niche might have led to societies of little more than the crowded poor after a very few generations. People might then have changed from ice-age hunters to an agricultural peasantry of the most depressed Asiatic type in an evolutionary instant, and would be essentially without hope of working the niche-changing magic a second time. But people were spared this because they learned to increase the size of the cake from which their niches would be cut, even as their numbers rose to bite out more and more slices.

With energy derived from things other than food or muscle, and by the use of materials and systems unknown to the primeval earth, people have been able to provide a larger cake of resources to share out among the niche slices. It seemed not to matter that their Darwinian family habits led to ever-rising numbers, because the new techniques created continually more resources for civilized niche-spaces. The cake that was divided among the people grew larger and larger. But these improvements went in stages, often reaching plateaus where ingenuity stagnated while the cake lost some crumbs. Rising numbers of people then caught up with the cake; the result was those periods of unrest that are the substance of history.

Wealth and an expanding way of life are possible so long as numbers are small compared with the total niche-space (or cake) made available by any level of technique. The satisfactory ways of life to which people aspired could be enjoyed by most so long as populations were not too large. But the breeding strategy has always worked to build ever-larger populations, and, as a consequence, there have always been rich and poor. Jesus of Nazareth is recorded as saying that the poor would be with us always. This cry of despair from a miserably crowded corner of the Roman Empire showed an awful understanding of a Darwinian breeding strategy retained by people who had escaped the other restraints of an animal way of doing things. Geometric increases in number can always overtake any increase in the size of the cake that technical ingenuity can devise. Perhaps ecology's first social law should be written "All poverty is caused by the continued growth of population."

In recent decades a few nations of the West have become so ingenious as to rush the size of the cake toward the upper limit set by such fixed boundaries as the amount of space available at the surface of the earth. Their populations have not yet caught up, so that they have, albeit on a purely local scale, reduced the numbers of people who must live in poverty. Moreover birth rates in these nations have fallen a little as people have changed their ideas about how many children they can afford. But populations rise nevertheless. A half of one percent compound interest is ample to take up the slack of resources in a few generations, then mass poverty will return. Relative poverty, of course, is with us all along.

While an expanding society has surplus resources, the better-off can concern themselves with raising the standards of the mass. But once the numbers of the people begin to catch up with the resources, the leaders will find that even their own ways of life are threatened. They must look to their own privileges, becoming a repressive ruling class.

Out of the twin pressures—rising numbers and a governing class seeking to defend an established and wider way of life—a caste system may be forged. In old India, people of high caste had ample lives, civilized and cultured, requiring many resources. Men of the lowest caste were at bare subsistence, and there were ranked castes in between. People did not know that the poverty in which most lived was caused by their breeding strategy, or if they did know they did not care. Surplus people in high castes could be relegated to castes below, where they took up less room. In the lowest caste, where most of the people were, each couple still raised the number of children they could afford. But they could not afford very many. The records of the first British conquerors of India show clearly that infanticide was common among the Indian peasants. This was the best method those of low caste could find to keep their family size down to what they could afford.

Another familiar caste system is that of the English in past centuries. This was less rigid than the Indian and more benign. The lowest caste of laborers lived ample lives by comparison with Indian peasants, at least they did before nineteenth-century industrialism, and there was comparatively little of nasty things like infanticide used to keep the family size in check. This was because the surplus people could be shipped to new lands overseas, lands with low populations of hunter-gathering peoples, which had been seized by British arms.

A ruling class that finds its way of life threatened by rising numbers will sense in a crude way that resources, particularly land for the younger sons, are in short supply. One obvious way to remedy this is to take resources from others by force. The taking of the New World from the aboriginal peoples by Europeans is an obvious example of such aggression, but the more spectacular wars of history fit these circumstances too. The conquests of the Great Captains are readily understood as the products of people's habit of changing niches but not breeding strategy.

Alexander is called "The Great" because he built an empire, destroying numerous armies and imposing a Greek way of life on the known world of his day. The achievement was too vast actually to be the work of one man. It can perhaps be partly explained by the superior technique and superior discipline of the Greek armies, coupled with the superior generalship of a clever young man educated by Aristotle. But the real explanation can be found in the history of Greece in the centuries before. These were years of strife, of technical advance, and of rising population. Greece established colonies, little overflow bits of Greece that should absorb her surplus people. That these should be expanded by ever larger wars was inevitable if the population continued to rise. And it would rise as long as every couple continued to raise as many children as it thought it could afford. Given that the state had developed advanced military techniques, a conqueror such as Alexander must come eventually. It is important to note that what was at stake for Greeks was not brute survival, not the ability to feed the people, but the Greek way of life. This was only possible within certain limits of density. All advancing societies have such critical limits to density. Once they are exceeded, and if there are weaker states nearby, they will go to war. Ecology's second social law might read "Agressive war is caused by the continual rise of population in rich societies."

The Greek Empire quickly disintegrated. The Roman Empire, which was built later and in a like manner, lasted longer, probably because many of the Roman conquests were of barbarian lands. The peoples of these had more primitive techniques that yielded subsistence living for fewer people, so their lands were underdeveloped by Roman standards and thus could absorb excess population for a long time. But the pressures of human crowds still caused extreme misery and poverty in the Roman Empire. It was a voice from those days that told us that the poor would be with us always.

Behind all the great aggressive conquests of history has been a rising population of people who have for a while hoped for a rising standard of life. They go to war to preserve that way of life by conquering fresh resources. But always their breeding strategy has been, not to choose a family size that will preserve forever the expanded way of life, but to choose the number of children they think they can afford. This means that population pressures within the empire they establish will destroy the very way of life it sought to maintain.

After the emergence of a city-state, an ecological model of human affairs predicts poverty, an upper class that becomes oppressive, caste systems, aggressive wars, empires, and the final dissolution of the empires as population catches up, producing a rebellious mass that cannot be governed indefinitely.

The historian Arnold Toynbee traced the rise and fall of all the civilizations for which we have records, and found a common pattern: civilizations arise in marginal lands. Toynbee says that the people need the spiritual shock of a hard environment to give of their best. An ecologist is not surprised to learn that the marginal lands foster aggressive civilizations for it is there that the pressure of rising numbers will be felt first, forcing expansionist zeal. From then on Toynbee's reconstruction is as predicted by the ecological model. There is for a time a "creative minority" of people whose example is followed by the mass. But the "creative minority" changes to a repressive "dominant minority," which the mass no longer emulates, becoming instead a sullen "internal proletariat." A "saviour with a sword" then builds a "universal state." But even this decays. The names are Toynbee's but what they represent are expected from an ecological analysis.

In the end, Toynbee notes that a world religion rises from the oppressed proletariat and persists long after the empire has fallen. An ecologist needs merely to note that much of the appeal of such world religions lies in their counsel to the oppressed to endure. "Nothing can be done; the poor are with us always; rely on your spiritual strength and make the best of things." People in a subsistence niche can do no more.

Nine thousand years have gone by in these cycles of history, each cycle the inevitable outcome of that initial event when people escaped from the animal condition of a fixed niche but did not modify their breeding strategy. Perhaps it is as well that we did not understand the role the people's choice in children was to play in our destiny, for most of the splendid things we have done have resulted from this as well as the hardships and miseries. With each stirring time of expansion we have gained beyond all reckoning in wisdom and the understanding of the life that is possible for us, and it has not all been lost with each subsequent collapse.

But now our populations are larger than any that have gone before, and we think our civilization is at least as good. Our ingenuity has made this possible by enlarging the cake of resources, both through improved agriculture and through fossil-fuel-driven industry. Not only have we in the West fed all our people but we have provided outlets or surrogates for many of the ice-age cravings. Travel and speed give the illusion of adventuring; moving images on screens offer the emotions effective placebos of the living for which we were made. The aspirations kindled when we began this epoch those nine millennia ago are now shared by more people than ever before. Yet this success will be short-lived if the population continues to grow. In a few more generations the troubles that have destroyed every civilization before ours will be upon us. Our way of life will be threatened by rising numbers, and we may expect some modern version of the revolutions and struggles that brought down our predecessors.

A likely consequence, or even prelude, to these struggles must be aggressive war. This means nuclear war. If this seems absurd, consider the plight of people in a technologically advanced island state. Poverty has long been in decline, but the aspirations of the people are high. Already the crowding in their island is such that the upper limits set by the pool of possible niches required for our incompressible requirements is reached. Privacy wanes. Rationing and restrictions must be placed on the use of open space in ways so stringent that the young find no outlet for the call to adventure that once adapted us to ice-age life. Rebellion and crime seem to increase, and a striving for resources to live well becomes disguised as a struggle for personal liberty. The excess energies of the people are for a time diverted overseas, particularly to the markets where they must sell their industry to pay for the raw materials and food that alone make a high standard of life possible in their crowded island. But the expansion overseas leads to friction, just as did the colonies of the Greek city-states. Rising demand for raw materials and food in the supplier continents then absorbs all the supply and there is none to ship away to islands. The island people, already restive from over-crowding, must relinquish some of their standard of life, some of what they see as liberty. The history of such aggressive island states as Britain or Japan gives little comfort to those who think that island people would not strike if faced with such a future.

Those who have brooded about the possibilities for nuclear war have concentrated on the chances of battle between the superpowers. Even ruthlessly objective analysts such as Herman Kahn have done this. But these continental states are very far from any ecological need to war. They have achieved high technological advance (a large cake) without dense populations. The Soviet Union owns one-sixth of the land surface of the globe and the United States is nearly as well off. Their fears of each other are trivial compared with the fears of poor states of the past faced with an emergent people led by a great captain. And the might of each superpower is such that they need never fear the aggression of some crowded beleaguered state desperately needing to expand. The superpowers do not have an interesting future for military theorists. For a nuclear war, a technologically advanced island and a backward continent are the most likely protagonists.

If numbers of people rise quickly all round, I take a nuclear war of aggression to be only a matter of time, of a few generations at most. But the rate of population increase is falling in all developed countries, and it is from one of these that a nuclear strike must come. Very much depends, therefore, on what is causing this fall in the rate of growth. We need to know if the cause is a fundamental one that will let the fall continue.

It is evident that we have not changed our ancient Darwinian breeding strategy of every couple choosing the number of children they think they can afford. Therefore the fall in birth and growth rates must be due to young couples changing their ideas on how many children they can manage. And this is clearly the truth of it. We live in broad niches, what we call "affluence." This requires many resources per person, and it requires many resources per child. Already the press of numbers on what we have is such that a young couple may be hard put to meet the requirements of their own affluence, so that winning more resources to provide the same broad niche for children does not seem easy to them. Moreover, educating children to all these satisfying ways takes time, and time is a parameter of niche that no ingenuity can stretch. In these circumstances no couple could expect to rear many children to be able to live up to the standards they themselves had come to expect, and a small family is an entirely predictable consequence of continuing to apply our Darwinian breeding strategy. People still have the largest possible families their resources allow, but their ambitions are such that their resources allow only a fraction more than the number that will replace them.

In this matter of the small family we have one strong advantage over all those past civilizations that have perished. They all accepted massive poverty, or slavery, or both, and this meant that there were always servants available to help with the rearing of children. The restraining influence of limited time was thus removed. A couple belonging to the dominant civilization could choose the size of their family with little worry about finding the resources to provide each child with the niche of the parents. Serfs would do the work; serfs could be displaced to provide other resources of the niche if need be. Although we have no data to test this conclusion, I think it certain that small families for the affluent is a phenomenon of our civilization alone. The others rushed headlong to their destruction.

People without affluence will never accept small families while the breeding strategy remains Darwinian, for it is the experience of affluence that is essential to those individual conclusions that the resources available will allow only a few children to be reared to the standards of the parents. Many people of public spirit have searched for ways to help the people of the bounding populations of poor countries to improve their lots. They have offered them birth-control devices. But modern birth-control methods allow us to perfect our Darwinian breeding strategy as never before. They give us the instrument to ensure that we have exactly the largest number of children we can rear to our agreed standard; not one too many and not one too few. The pill, the condom, and the intrauterine device are the most potent instruments for ensuring that the largest possible number of young people are recruited into the next generation. This is one reason why the populations continue to bound in the undeveloped lands.

So the patterns of family size in the modern world are fully understandable from ecological theory. They are, indeed, predictable consequences of the way our species has gone about living these nine thousand years. We must gauge the future of the next few generations of people on the assumption that these patterns will continue. Families will remain, with some fluctuation, small in the wealthy countries. Families will remain large in the poor.

The fact that the growth of affluent populations will remain slow must lessen the chances for a nuclear strike in the decades immediately ahead although there may be minor nuclear bickerings between India and China. Furthermore, by a great accident of history, the most formidable armaments are in the hands of two continental superpowers whose populations are not dense and whose ecological needs are not desperate. These powers can use their terrible weapons to bully any potential island aggressor out of its plans to attack some weak continent. Whatever happens at SALT (Strategic Arms Limitation Talks), it seems likely that both superstates will keep enough of their weapons to be able to play the bully indefinitely. And it also seems likely that self-interest will let them continue to accept that role. As populations mount only slowly, and while the superpowers keep their awful weapons, we can expect a long crowded future without the depopulating relief of a nuclear aggression.

This lets us read the future rather clearly. The vision is not good. It includes all the troubles of overcrowding that broke the empires of the past, but without the aggressive wars in which crowded peoples have often found relief. If our numbers stopped growing now, the ingenuity of advanced Western peoples is probably good enough to meet their aspirations for some time to come. But this means the two-child family as the norm in all Western communities, a norm established and maintained for generations. A new messiah might bring this about, but the prudent will not expect it to happen. So the future holds the prospect of slowly rising numbers of people, with rising aspirations, and without much hope of a successful aggression as a way out for them.

Technology can probably find raw materials and even energy for human manufactures almost without limit. In this sense the resources of a broad niche can be made to grow in step with the population. But other resources cannot be stretched much further. These resources provide space, privacy, some taste of adventure for the young, and the right to do sometimes as one pleases. These resources will have to be rationed. To do this will require more government and more bureaucracy. In countries with good government, fair shares will be had by all; in other countries, satisfying shares will be won by the few and subsistence for the rest.

We are about to crowd more and more people into our societies. We can feed them, clothe them, and shelter them. For a time at least we are going to deny them the right to aggressive war (or free them from it depending on your point of view). But we are surely going to force very many of them to live in niches that are not congenial to them. If we really would know what the future will be like, we need, therefore, some satisfactory definition of the kind of human niche we are about to deny to so many. I suggest that the work of philosophers for centuries has given us an understanding of what a desirable human niche must be. It was written down most clearly for us two hundred years ago in America by a group of literate men who thought profoundly about it, even as they fought for the right of their people to have it. We may say that a satisfying human niche is bounded by a set of unalienable rights, among which are life, liberty, and the pursuit of happiness. Our technology will continue to grant life. It is the other parameters of our niche that will be denied as our populations slowly crowd.

The future therefore holds ever greater restrictions on individual freedom. We will not be able to live as our fathers lived, and our traditional ways of doing things will seem like poems of the past. Nor will we be able to thrill to the voices of Great Captains urging us to take up our weapons in quest of liberty outside our borders. Liberty will fall progressively as the numbers rise, and obedient compliance with the majority will must take the place of individual initiative.

# Postlude

THE sun has glowed through the earth's skin of atmosphere and onto its rocky crust for four and a half thousand million years. It drove heat engines that worked by the power of running water, and by freezing and thawing so that it rendered the surface of the crust even as it writhed. It moved the gases of the atmosphere in perpetual cycles, even as it was the fuel for gradual chemical changes. There was no free oxygen at first, for this combined with iron, calcium, sulphur, and other elements as fast as it could be made. But the sun split oxygen from water high up toward the edge of the skin, which let the light hydrogen drift away. More importantly the sun came to be used by the first bacteria and blue-green algae that manipulated carbon to make fuel and discarded the oxygen with which that carbon was combined. Within two or three thousand million years they had pumped out so much oxygen that the atmosphere had acquired 20 percent of it. The atmosphere was then what we now call air.

As the oxygen reservoir built up, natural selection favored successions of life-forms that could tolerate and use the new surroundings. There was plenty of time for this working out of biological chemistry, certainly a thousand million years. At the end of the time, plants had been made that could undertake photosynthesis, and that could then use the free oxygen to burn the resulting sugars at will, releasing energy when they needed it. But free oxygen meant that some of the living things could burn plant sugars without having to go to the trouble to make them first. These others were the first animals.

When animals crop plants, there is an advantage in being a plant that cannot be cropped. New kinds of plants that held awkward chemicals, or that dispersed to remote places were preserved by natural selection. A start had been made toward that pattern of many species living side by side that is both the oddity and the richness of contemporary vegetation. Natural selection favored fresh strains of animal that were equipped to pursue each plant novelty, so that more kinds of animal appeared. Eventually some strains of these new animals could make a living not just by hunting plants, but by pursuing some of the array of plant eaters instead. These were the first carnivores. And they forced natural selection to preserve herbivores that had defenses against them. The first armament began to be carried by animals, and then hunter and hunted began an evolutionary race for large size, because large size is a defense that favors corresponding large size in the attacker. This was a very recent happening. We date it at about five hundred million years ago because that is when our fossil record of an array of animals with skeletons big enough to be seen with the naked eye first appears in the rocks.

The earth was then essentially the earth we know. Plants, their hunters, and the hunters of their hunters lived in air we would be able to breathe or in oceans whose saltiness we would recognize. On the patchy surface of this earth, the plants, together with their stacked-up set of hunting animals, were directed into local varieties as local physical circumstance influenced the endless games of hide and seek, even as their endless motions tended to mix these diverging populations. Where the types were mixed, natural selection favored those with distinct ways that would let them go about the business of collecting energy and raising babies without wasteful strife. Those who bred in the greater peace were the most fit. Their descendants became new species, each being provided with a niche that let it peacefully coexist with its neighbors.

The business of every species was that of raising its own young, but this was always hard. Natural selection forced each individual to compete with its relatives for the food energy it would process into offspring. Many species were driven to the gambler's expedient of dividing their capital into tiny chips that would be available to cover every eventuality in an uncertain physical world; and there came to be hosts of tiny eggs and seeds. Others were given the technique of seeking a large return on capital by passing all the food they won into a few large babies. Neither of these extremes of breeding strategy made any difference to the eventual sizes of populations because numbers were set by the opportunities provided for each way of life on a finite earth.

The peaceful coexistence that natural selection forced on all these animals and plants made them share the living places. They had common interest in the essential raw materials of these places, in phosphorus, potassium and the rest, and these common interests modified the purely physical cycles by which these things were moved around. There was a biological component to the natural cycling that we recognize when we talk of an ecosystem.

But the energy drive of the living things remains small compared with that of the inanimate world. Green plants convert solar energy with an average efficiency of less than 2 percent, because of shortage of carbon dioxide and other raw materials. More than 98 percent of the free energy falling on the earth, therefore, goes to the physical drives of ecosystems. All life reacts to this reality, accommodating to the physical world rather than molding it.

We see this accommodation to solar energy and earthly rocks in many of the grander patterns of nature; in the different shapes of plants in different countries, in the desert unproductivenes of the blue oceans, in the rarity of the large and fierce, in the changes in the state of water as we pollute or unpollute a lake, in the limits to food or the possibilities for human happiness. And this accommodation is even felt in degrees of the stability or balance in nature that seems so important a part of life on earth. Stability and balance are not so much functions of life acting on life as they are reflections of the underlying stability of physical systems. Perhaps the greatest error recurrent in ecological thought is that which claims stability as a function of biological complexity. The idea that species collect together or accumulate in ways that lead to stable entities is as old as ecology, but still without objective foundations.

Although natural selection has muted the competition between individuals of different species they must always struggle with individuals of their own kind for the necessities of life. This follows directly from the necessity of raising the most young possible. But even this competition can be muted in surprising ways, though only when there is a clear advantage to each individual of the compact. Territorial animals respect the neighborhood of another of their kind because there is more personal chance of fitness in yielding a claim and trying elsewhere than in pursuing a struggle from which little fitness can be won. The resulting patterns of ordered behavior are responsible for many of our feelings about a satisfying regulation of life in nature.

Yet the harsh facts of physical reality, fueled by that 98 percent of the free energy of the biosphere, threaten all species with accidents that can lead to extinction. As new species are continually forged by the random overlapping of populations and elimination of those who compete too strongly, other species are being removed from the inventory. The number of species living at any time and place is set by the balance between these two processes, the creation of the new and the removal of the old.

As long as the finding of new ways in which to live was left to natural selection, there was always a tenuous peaceful coexistence of the living things on earth. But eventually one kind of animal found it possible to keep occupying new niches at will, always adding the niche-spaces of others to its own, escaping the ancient constraint of a fixed niche that is imposed on all others by natural selection. This animal yet continued to obey the other dictum of natural selection, which is to raise the largest possible number of offspring. The activities of this new form of animal are inevitably hostile to the interest of almost all the other kinds, for it engages in aggressive competition, instead of peaceful coexistence, in its drive for more and more young. It has been carrying on this new way of life for only nine thousand years.

# Ecological Reading

THE ecology of the professionals has been slow to emerge from the primary literature in specialist journals, and their subject has tended to be reported by writers who have peered in on their activities from outside. Any public library will have a thick collection of cards filed under "ecology," but the cards will not list many titles that discuss the subjects in this book. When they do touch on our common ground it will be, often enough, to express ideas that are claimed to be ecological and that I have tried to show are in error; that the atmosphere is at risk, that simplifying ecosystems makes them unstable, that we ought to farm the oceans. It may be that the public library has nothing on our sort of ecology at all. If it does, it may have Odum's textbook. This, at least, is good.

Before 1970 there were two general texts on ecology that have left a permanent mark, both of which can still be read with profit. The first was Charles Elton's *Animal Ecology*, first published in 1927, and then in various editions with only minor changes. It is to this book more than any other that we owe the universal recognition of food webs and food chains. It is also a Darwinian book, full of the true insights of the evolutionary biologist. It fueled many in the profession who made the modern advances. It is literate and good reading. The second was E. P. Odum's *Fundamentals of Ecology*, first appearing in 1953 and greatly revised in later editions. This is the book that made the ecosystem fashionable. Odum made himself the spokesman to the college community for the ideas of energy flux through ecosystems, which had been developed by Lindeman and Hutchinson at Yale. "The Odum" held up ecology as a real science for fifteen years while many another book treated it as nature-watching. The latest edition (1971) is still a very good source of sound ecological information, though some of us think it stresses the growth of ecosystems through succession a mite too strongly.

In the fifteen years that Odum held almost undisputed sway, many who had got part of their training from his work were both researching and developing their own syntheses. They were still mostly silent outside their own journals even as the environmental crisis broke and many regrettable things were said by others in the name of ecology. Then their textbooks came out in a rush in the early seventies, the peak of the wave being the six months crossing the new year of 1973 when five new texts suitable for beginning college students appeared, all of them books with the modern professional emphasis. Together with Odum they provide the best source of solid ecological information to be gleaned without previous training. Krebs, McNaughton and Wolf, and Collier *et al.* have something of the usual pedagogic aura of books you teach from; both Rickleffs and Colinvaux have tried more for textbooks that are meant to be read.

The naturalist writer, Aldo Leopold, had great importance in the development of the subject, particularly with his *Game Management* of 1936. In reading Leopold, however, it is necessary to note that he did much to cause the overemphasis on the controlling effect of large predators on their prey that is still current. His account of the supposed population explosion of deer on the

Kaibab Plateau of Arizona after the predators were shot has recently been shown to be unfounded (Caughley, 1970), though it appears in even some of the latest books such as that of Rickleffs.

Two books important for establishing rival positions in an ecological controversy appeared in 1954, both works of the literary power in argument that gives pleasure to the scholar. David Lack, with his *Natural Regulation of Animal Numbers*, developed the Lotka-Volterra-Gause competition models into a general theory for the density-dependent control of all animal numbers, drawing his data largely from studies of birds. Andrewartha and Birch, in their *Distribution and Abundance of Animals*, argued the opposite; that the competition models had little relevance to real life and that random hostilities of weather, coupled with widespread dispersal, set the numbers actually living. In the twenty years since these books appeared most ecologists have absorbed the best parts of both into their consciousness.

Three books of essays by G. E. Hutchinson bring the crux of ecological arguments out from under the equations and jargon with which, alas, most of our thoughts are clothed. These are *The Itinerant Ivory Tower, The Enchanted Voyage,* and *The Ecological Theater and the Evolutionary Play.* Modern ecology has nothing else as literate to offer, though MacArthur's *Geographical Ecology* sets down the later ecological ideas in an accessible way.

There is very little literature on the theme of my last chapter, "The People's Place." I have elaborated the historical model in my new book, *The Fates of Nations,* which Simon and Schuster has scheduled to appear in the spring of 1980. My three other published accounts I cite. The nearest attempt I know of to think in parallel lines is Heilbroner's *The Human Prospect,* but this is an economist talking. The book has the flawed ecology that accepts arguments like the one that says that simplifying ecosystems makes them unstable, having been written before the refutations of these ideas got fairly out of our invisible college. Heilbroner also thinks nuclear war a possibility, but has the aggressor as the backward continent instead of the other way round, which my ecological model suggests. The best book to describe the evolutionary basis of the human predicament is that by Rozensweig. The best essay is Hardin's "Tragedy of the Commons."

The rest of the titles in the bibliography that follows are references for the discussions in this book. Where I have mentioned an ecologist's name in the discussions I cite the reference below. References for the more general ideas can be found in any of the six texts described in the third paragraph above. Other references are included as follows. For productivity, its measurement, and for gradient analysis; Whittaker (1975). For schools of plant sociology; Whittaker (1962) and Oosting. For complexity stability theory and its refutation; MacArthur (1955), May, and Goodman. For the effect of grazing animals; Harper. For measures of efficiency of energy conversions; Slobodkin. For Australian magpies; Carrick. For the Maine gunners; Stewart and Aldrich. For the vedalia beetle story and other examples of biological control; De Bach. For character displacement; Brown and Wilson. For the hypothesis that lakes grow fertile with aging; Deevey and Livingstone. For the chemistry of the oceans and other geochemistry; Garrels and McKenzie. For the energy flux in the biosphere; Morowitz.

Andrewartha, H. G., and L. C. Birch, 1954. *The Distribution and Abundance of Animals*. Chicago: University of Chicago Press, p. 782.

Broeker, W. S., 1970. "Man's Oxygen Reserves," *Science*, 168: 1537-1538.

Brown, L. L. and E. O. Wilson, 1956. "Character Displacement," *Systematic Zoology*, 5: 49-64.

Bryson, R. A., 1966. "Air Masses, Stream Lines, and the Boreal Forest," *Geographical Bulletin*, 8: 228-269.

Carrick, R., 1963. "Ecological Significance of Territory in the Australian Magpie, *Proceedings of XIII International Ornithological Congress*, 9: 740-753.

Clements, F. E., 1916. *Plant Succession: An Analysis of the Development of Vegetation*. Carnegie Institution of Washington Publication 242, facsimile reprint by Haffner.

Collier, B. D., G. W. Cox, A. W. Johnson, and P. C. Miller, 1973. *Dynamic Ecology*. Englewood Cliffs, N.J.: Prentice-Hall, p. 563.

Colinvaux, P. A., 1973. *Introduction to Ecology*. New York: John Wiley and Sons, p. 621.

————, 1975. "An Ecologist's View of History," *Yale Review*, 64: 357-369.

————, 1976. "The Human Breeding Strategy," *Nature*, 261: 356-357.

————, 1976. "The Coming Climactic," *Bulletin of Ecology*, 56: 11-14, and in "The American Years," The Massachusetts Audubon Society, Lincoln, Mass., 1976.

De Bach, P. (ed.), 1964. *Biological Control of Insect Pests and Weeds*. New York: Reinhold, p. 844.

Deevey, E. S., 1942. "Studies on Connecticut Lake Sediments III. The Biostratonomy of Linsley Pond," *American Journal of Science*, 240: 235-264, 313-324.

————, 1955. "The Obliteration of the Hypolimnion," *Mem. Ist. Ital. Idrobiol.*, suppl. 8: 9-38.

Elton, C. S., 1927. *Animal Ecology*. New York: Macmillan, p. 209.

Garrels, R. M., and F. T. McKenzie, 1971. *Evolution of Sedimentary Rocks.* New York: W. W. Norton, p. 397.

Gates, D. M., 1965. "Heat Transfer in Plants," *Scientific American,* 213: 76-86.

————, 1968. "Energy Exchange between Organisms and Environment," *Australian Journal of Science,* 31: 67-74.

Gause, G. F., 1934. *The Struggle for Existence.* Baltimore: Williams and Wilkins, p. 163.

Goodman, D., 1975. "The Theory of Diversity and Stability in Ecology," *Quarterly Review of Biology,* 50: 237-266.

Hardin, G., 1968. "The Tragedy of the Commons," *Science,* 162: 1243-1248.

Harper, J. L., 1969. "The Role of Predation in Vegetational Diversity," Brookhaven Symposium in Biology No. 22, *Diversity and Stability in Ecological Systems,* pp. 48-62.

Heilbroner, R. L., 1974. *An Inquiry into the Human Prospect.* New York: W. W. Norton, p. 150.

Horn, H. S., 1971. *The Adaptive Geometry of Trees.* Princeton, N.J.: Princeton University Press, p. 265.

Hornocker, M. G., 1969. "Winter Territoriality in Mountain Lions," *Journal of Wildlife Management,* 33: 457-464.

Howard, H. E., 1920. *Territory in Bird Life.* New York: E. P. Dutton, p. 308.

Hutchinson, G. E., 1953. *The Itinerant Ivory Tower.* New Haven, Conn.: Yale University Press, p. 261.

————, 1962. *The Enchanted Voyage.* New Haven, Conn.: Yale University Press, p. 163.

————, 1965. *The Ecological Theater and the Evolutionary Play.* New Haven, Conn.: Yale University Press, p. 139.

Janzen, D. H., 1970. "Hervibores and the Number of Tree Species in Tropical Forests," *American Naturalist,* 104: 501-528.

Klopfer, P. H., 1969. *Habitats and Territories.* New York: Basic Books, p. 117.

Krebs, C. J., 1972. *Ecology: The Experimental Analysis of Distribution and Abundance.* New York: Harper and Row, p. 694.

Lack, D. L., 1954. *The Natural Regulation of Animal Numbers*. New York: Oxford University Press, p. 343.

Leopold, A., 1933. *Game Management*. New York: Charles Scribner's Sons.

Lindeman, R. L., 1942. "The Trophic Dynamic Aspects of Ecology," *Ecology*, 23: 399-418.

Livingstone, D. A., 1957. "On the Sigmoid Growth Phase of Linsley Pond," *American Journal of Science*, 255: 364-373.

MacArthur, R. H., 1955. "Fluctuations of Animal Populations, and a Measure of Community Stability," *Ecology*, 36: 533-536.

————, 1958. "Population Ecology of Some Warblers of Northeastern Coniferous Forests," *Ecology*, 39: 599-619.

May, R. M., 1973. *Stability and Complexity in Model Ecosystems*. Princeton, N.J.: Princeton University Press, p. 265.

McNaughton, S. J., and L. L. Wolfe, 1973. *General Ecology*. New York: Holt, Rinehart and Winston, p. 710.

Mech, L. D., 1966. "The Wolves of Isle Royale," *Fauna of the National Parks of the United States*, Fauna Series 7, U.S. Government Printing Office, Washington, p. 210.

Morowitz, H. J., 1968. *Energy Flow in Biology*. New York: Academic Press, p. 179.

Murie, A., 1944. "The Wolves of Mount McKinley," *Fauna of the National Parks of the United States*, Fauna Series 5, U.S. Government Printing Office, Washington, p. 238.

Odum, E. P., 1971. *Fundamentals of Ecology*. 3rd edition, Philadelphia: W. B. Saunders, p. 574.

Oosting, H. J., 1956. *The Study of Plant Communities*. 2nd edition, San Francisco: W. H. Freeman, p. 440.

Owen-Smith, N., 1971. "Territoriality in the White Rhinoceros *(Ceratotherium simum)* Burchell," *Nature*, 231: 294-296.

Petersen, R., 1975. "The Paradox of the Plankton: An Equilibrium Hypothesis," *American Naturalist*, 190: 35-49.

Rickleffs, R. E., 1973. *Ecology*. Newton, Mass.: Chiron Press, p. 861.

Rosenzweig, M. L., 1974. *And Replenish the Earth.* New York: Harper and Row, p. 304.

Sanders, H. L., 1968. "Marine Benthic Diversity: A Comparative Study," *American Naturalist,* 102: 243-282.

Schaller, G. B., 1967. *The Deer and the Tiger.* Chicago: Chicago University Press, p. 370.

Shannon, C. E., and W. Weaver, 1949. "The Mathematical Theory of Communication," Urbana: University of Illinois Press.

Slobodkin, L. B., 1962. "Energy in Animal Ecology," *Advances in Ecology,* 4: 69-101.

Stewart, R. E., and J. W. Aldrich, 1951. "Removal and Population of Breeding Birds in a Spruce-Fir Forest Community," *Auk,* 68: 471-482.

Tansley, A. G., 1935. "The Use and Abuse of Vegetational Concepts and Terms," *Ecology,* 16: 284-307.

Transeau, E. N., 1926. "The Accumulation of Energy by Plants," *Ohio Journal of Science,* 26: 1-10.

Whittaker, R. H., 1962. "Classification of Natural Communities," *Botanical Review,* 28: 1-239.

———, 1975. *Communities and Ecosystems.* 2nd edition, New York: Macmillan, p. 385.

# Index

abundance, relative: question of, 7, 11 (*see also* numbers problem); discussion, 199-211; dominance in plants, 63-64, 127-130; large animals, 18-31; Shannon-Wiener measure, 201-202

agriculture: efficiency of crops, 34-36; efficiency of herbivores, 46; green revolution plants, 45; monoculture, 203; nutrient cycles, 74-81; oceans, 89-92, 96; temperate north, 80; tropics, 74. *See also* pasture

aging, of ecosystems, 114-116

air pollution, 97-107

Alaska pipeline, 208

Alexander the Great, 224-225

algae: efficiency of, 41-42; productivity, 41; smallness in open sea, 23, 83-88; in microcosms, 130-132; diversity in plankton, 188-189

algal culture, 41-42

allelopathy, *see* chemical defenses

Amazonian highway, 209

arctic: food chains, 19-21; trees absent, 55-57; front and treeline, 60-61; diversity in, 195-196; stability of, 203-205

associations, 64-72; Zurich-Montpelier definition, 65-66; classification of, 66-68; of Uppsala school, 68; on mountain sides, 70-71; as loose symbiosis, 72

atmosphere: carbon dioxide regulation, 103-107; composition of, 97-107; indestructibility, 103; oxygen and nitrogen maintenance, 100-103; regulated by life, 100

Australian magpies, group territories of, 176-177

balance of nature: constant numbers problem, 8; crowding, density-dependence and competition, 136-149; role of predators, 150-161; role of territory, 162-182; complexity-stability theory, 199-211; explained, 209-210

biological control of pests, 157-160

biomass, 24

biome, *see* formations of plants (a term describing same units but which includes animals in descriptions)

**Library of Congress Cataloging in Publication Data**

Colinvaux, Paul A., 1930-
    Why big fierce animals are rare.

    Bibliography: p.
    Includes index.
    1. Ecology.    I.    Title.
QH541.C64        574.5        77-71977
ISBN 0-691-08194-8
ISBN 0-691-02364-6 pbk.